IMAGES
of America

ARIZONA-SONORA DESERT MUSEUM

On the Cover: The original animal line at the Arizona-Sonora Desert Museum is seen from the vantage point of the black bear enclosure in the 1950s.

IMAGES
of America

ARIZONA-SONORA DESERT MUSEUM

Peggy Pickering Larson
and William Ascarza

ARCADIA
PUBLISHING

ISBN 978-0-7385-8671-7

Published by Arcadia Publishing
Charleston, South Carolina

Printed in the United States of America

Library of Congress Control Number: 2010910134

For all general information, please contact Arcadia Publishing:
Telephone 843-853-2070
Fax 843-853-0044
E-mail sales@arcadiapublishing.com
For customer service and orders:
Toll-Free 1-888-313-2665

Visit us on the Internet at www.arcadiapublishing.com

For Merv, who took animals out of cages, and for Lane and Lynn, who grew up in the museum's shadow
—Peggy Pickering Larson

To the museum founders and early employees who pioneered public interpretation of the Sonoran Desert a living reality
—William Ascarza

Contents

ACKNOWLEDGMENTS

Unless otherwise noted, all images appear courtesy of the Arizona-Sonora Desert Museum. We would like to thank the following individuals who have contributed to this book: Emmanuel Ascarza, Marilyn Ascarza, Bob Brown, Wendall Brown, Dr. Rick Brusca, Hunter Byrd, Bob and Uta Campbell, George Carpenter, Roger E. Carpenter, Lorie Cavalli, Dr. Debbie Colodner, Greg Crooks, Joseph Curamen, Dr. Mark Dimmitt, Kenny Don, Robert Edison, Sloane Haywood, Bill Hornbaker, Jared Jackson, Karen Krebbs, Lynn Larson, Mervin W. Larson, Mary Powell McConnell, Carol Madeheim, George Montgomery, Sam Negri, Camille Pons, Beth Rice, Gareth Rixton, Hector Rodriguez, Tim Rogers, Jeanne Rosengran, Nancy Serensky, Tim Vimmerstedt, Elizabeth Woodin, William H. Woodin, and Barb Yates.

INTRODUCTION

The Arizona-Sonora Desert Museum is one of the most renowned institutions in the United States. Hosting half a million visitors a year, the Desert Museum has been rated as one of the top 10 museums in the world. Spanning nearly six decades, the museum has fostered respect and appreciation for the wildlife and vegetation of the Sonoran Desert. From its beginnings in 1952, the museum has pioneered the use of naturalistic habitats for its animals and interpreted in great depth just one particular ecosystem—the desert. Over the years, it has evolved into a major, premier institution, the zoo that zoo directors talk about.

The Arizona-Sonora Desert Museum began as a concept in the mind of New York native William H. Carr. A former director of the Bear Mountain Trailside Museum in New York, Carr arrived in Tucson in 1944. He immediately saw the need to establish a similar museum that would focus on interpreting native desert plants and animals. Needing funding and support for his ambitious vision, he convinced the renowned conservationist and philanthropist Arthur N. Pack to contribute funds from his family's foundation. Carr also solicited the support of the Pima County Park Committee, whose first chairman, C. B. Brown, was also the founder of Tucson Mountain Park. These connections helped establish the Arizona-Sonora Desert Trailside Museum, amid some local Tucson residents' opposition to an idea they feared might be equivalent to a roadside snake farm.

A second challenge was to obtain a location for the museum. This was resolved through an agreement between the Pima County supervisors and the museum's founding fathers. The Tucson Mountain House, built by the Civilian Conservation Corps, and its 98 acres of land 12 miles west of Tucson in Tucson Mountain Park met the need, and the county-owned property was leased to the museum. The Arizona-Sonora Desert Trailside Museum opened on Labor Day 1952 to an enthusiastic crowd of Tucsonans who desired to learn more about their local ecosystem. Shortly thereafter, the word "trailside" was removed from the museum's title.

Throughout the 1950s, Carr's vision of a living museum continued to grow with the addition of more staff and exhibits. Both William H. Woodin and Mervin W. Larson, future directors of the museum, served as early staff members. Their talent for innovation and organization proved invaluable to the museum. Additional early staff included Lew Walker, Hal Gras, and Merritt Keasey. A board of trustees was established with members that included naturalist author Joseph Wood Krutch and Roy Chapman Andrews, a renowned paleontologist.

Early interpretations for which the museum was famous included Water Street, which educated visitors on the importance of water conservation in the desert. Other exhibits included the Tunnel, built in 1957 to interpret life underground in the desert, and the amphibian room opened in 1958. A personable early feline resident, George L. Mountainlion, served as a public relations ambassador for the museum. A favorite among visitors, he became the official icon of the museum and also wrote a weekly column (with human assistance) about museum happenings for the local newspaper.

The 1960s continued in the tradition set the previous decade, with new exhibits opening along with increased international acclaim. Animal enclosures became more natural as the museum refined techniques for exhibition. A riparian habitat including a beaver and otter exhibit was initiated in the latter half of the decade. This was followed by the opening of the bighorn sheep exhibit with massive artificial rockwork molded into steep cliffs and rock tanks. Museum personnel were involved in conservation and scientific studies relating to Mexico and were influential in the Mexican government protecting islands in the Gulf of California, beginning with Isla Rasa in 1964. A demonstration desert garden emphasizing practical water usage was implemented in three phases beginning in 1967, helping to promote native plant gardening among the local population.

The 1970s brought about an increase in public education and outreach toward conservation of land and species protection with neighboring Mexico. Museum personnel acted as consultants, establishing an institution similar to the Desert Museum in Hermosillo, Sonora, Mexico, the Centro Ecológico. The docent program was initiated in 1972. The creation of the Cat Canyon exhibit and the Earth Sciences Center were other hallmarks of the 1970s.

An earth history and mining gallery opened in 1981 at the Earth Sciences Center. By the middle of the decade, the Mountain Woodlands Exhibit was built, replacing the previous outmoded exhibits for mountain lions, black bears, Mexican gray wolves, and deer. A new technology called Invisinet was patented by the Arizona-Sonora Desert Museum for the purpose of fencing live animal enclosures. A volunteer program apart from the docent program was established during this period.

The 1990s ushered in a new era of exhibits and programs. The hummingbird aviary, desert grasslands, Desert Loop Trail, javelina exhibit, coati exhibit, and pollination gardens were completed. New programs included tortoise adoption, migratory pollinators, and environmental enrichment for museum animals. The establishment of the Arizona-Sonora Desert Museum Press and the Art Institute were additional accomplishments in the 1990s. A new restaurant and gift shop opened during this time.

From 2000 to the present, the Arizona-Sonora Desert Museum has branched out further with exhibits and programs. The Ancient Arizona Exhibit resulted from the discovery of the sonorasaurus fossil in the Whetstone Mountains south of Tucson. Life on the Rocks opened in 2008, depicting the many animal species that inhabit the rocky slopes of our "sky islands" or mountain ranges. The Center for Sonoran Desert Studies conducts research into topics including bats, endangered species in the Gulf of California, and invasive species of vegetation, such as bufflegrass. Educational programs such as Raptor Free Flight, Live and (sort of) on the Loose, Running Wild, and Earth Camp have enhanced visitor turnout and education. The development of the Arizona-Sonora Desert Museum Digital Library has established a world-class authoritative file of images and information pertaining to the Sonoran Desert.

The Arizona-Sonora Desert Museum will undoubtedly remain one of the premier zoos and botanical gardens in the world throughout the 21st century. The same pioneering effort of its forefathers remains evident in the quality, hard work, determination, and innovation of its scions.

One

1950s

With some large mammals outside and small mammals, reptiles, and a few minerals inside, the Arizona-Sonora Desert Museum opened Labor Day 1952. Located in the desert 12 miles from the nearest urban area, the museum was reached by a rutted dirt road over a desert mountain pass. Visitor amenities were limited. On opening day, a dusty parking lot overflowed with cars, restrooms were limited to the generic two, only one drinking fountain was available (with the temperature exceeding 100 degrees), and a food and drink concession was lacking. There were some local skeptics who predicted an early demise for the undertaking.

They were wrong. The museum's spectacular setting was centered around the handsome Tucson Mountain House, located in the finest saguaro cactus forest in the world and backed by rugged mountains. The cofounders were visionaries who gathered around them a small, energetic, dedicated, creative staff and a supportive board. Through the efforts of these individuals and enthusiastic community support, the museum began its rapid development to becoming a world-renowned institution.

In the period from 1952 to 1960, the museum made spectacular progress in the construction of new exhibits and development of entirely new techniques for building them. In the Tunnel, the usual concept of viewing animals above ground was replaced by observing them in naturalistic underground retreats. The watershed exhibition demonstrated the use and misuse of water as a valuable commodity in the desert.

In the amphibian room, large-scale dioramas combining museum and zoo techniques depicted life in aquatic habitats. A wildlife photography blind constructed near a waterhole attracted wildlife and avid photographers. The *Arizona Desert Trails* television show aired weekly. The Desert Ark (a station wagon) brought animals, fun, and natural history education to schools and other groups.

By 1958, the museum received its one-millionth visitor and welcomed holders of its 1,000th membership. In the single month of January 1958, readers learned of the museum in *National Geographic*, *Sunset*, *Nature*, *Chicago Sunday Tribune*, and *San Diego Union*. And the ever faithful George L., the museum's literary lion, kept local folk informed of museum happenings in his weekly column for the *Arizona Daily Star*.

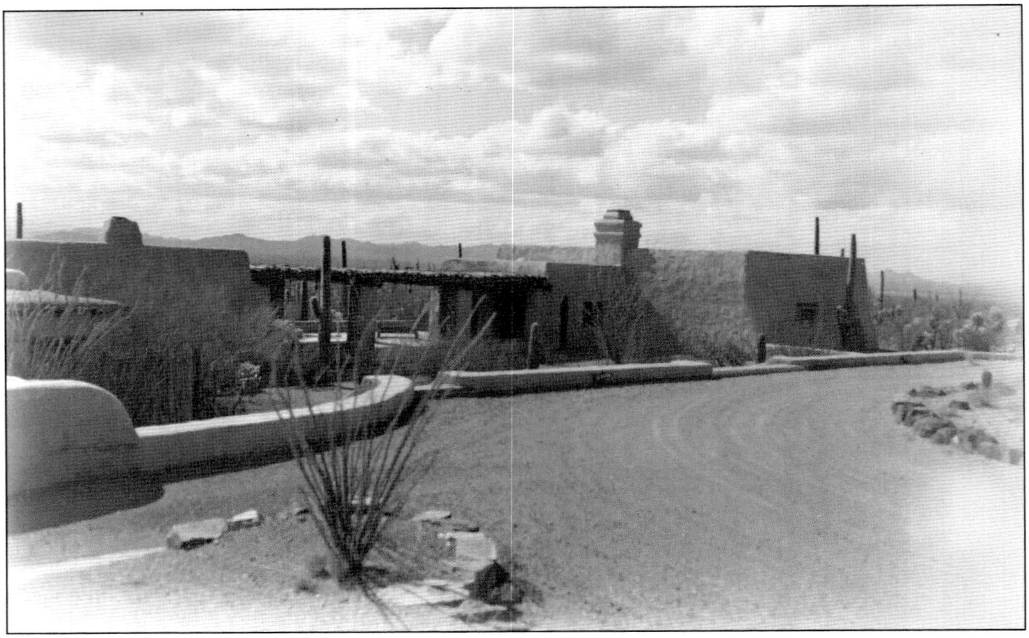

In the late 1930s, a complex of buildings, including the large Tucson Mountain House, was constructed in Tucson Mountain Park by the Civilian Conservation Corps (CCC) to serve as a headquarters for several southwestern parks administered by the National Park Service. The buildings were never used for that purpose and were first occupied by George and Lillian Adams, who established a restaurant there. Gas rationing during World War II soon made this impractical, and the buildings sat vacant. After the war, George and Alberta Zoellner lived in the buildings as caretakers, and the facility was sometimes used by picnickers, campers, and church groups. The Zoellners left when Bill Carr received permission to use the buildings for the Desert Museum. By then, pack rats, time, and termites had caused serious damage to the buildings, but once repaired, these and the saguaro forest surrounding them proved an ideal setting for the new Arizona-Sonora Desert Museum. (Below, courtesy of Zoellner family.)

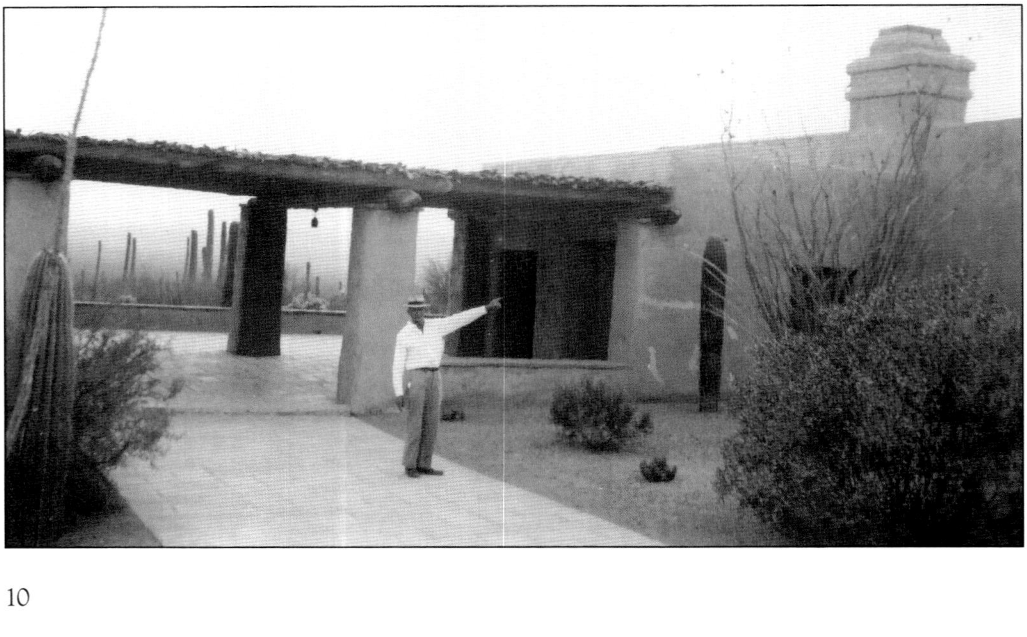

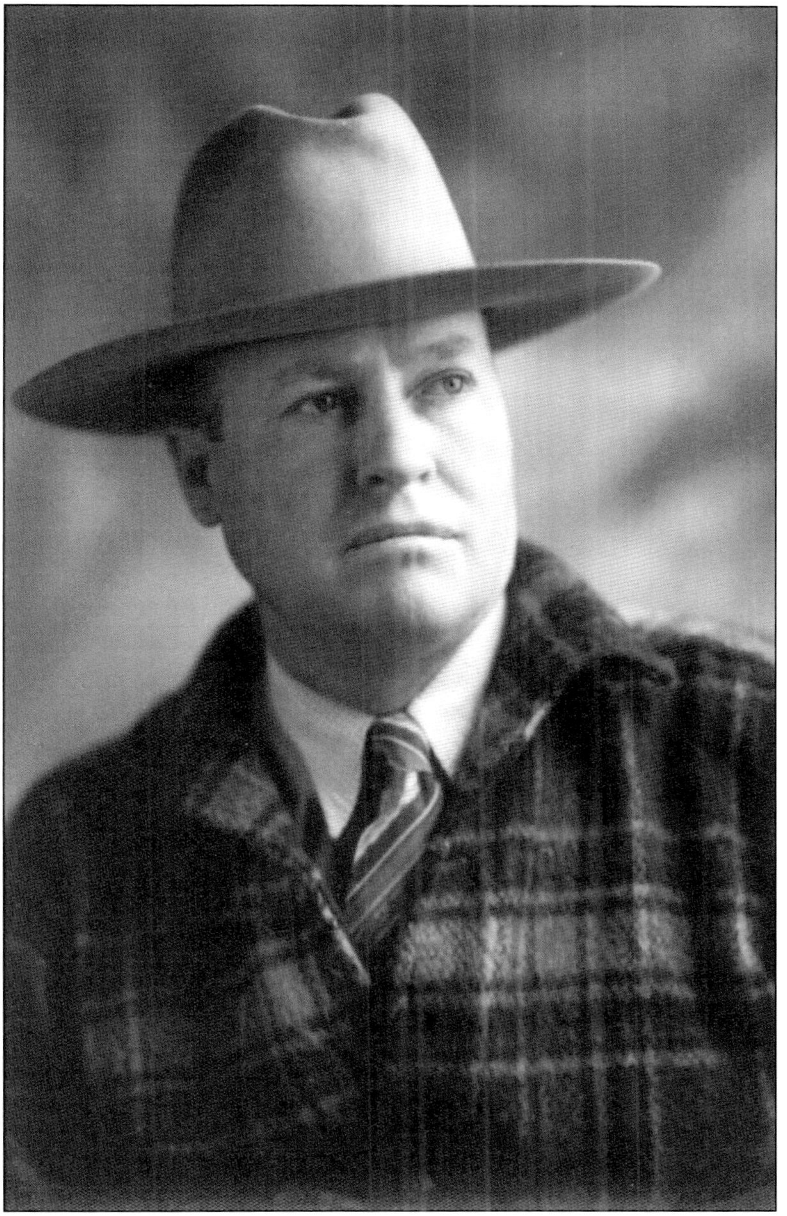

Cornelius B. "C. B." Brown arrived in Tucson in 1920 to serve as a Pima County agricultural agent and soon made an emotional connection to the desert landscape. He wrote, "The scenic qualities, luxuriance and variety of desert flora, abundance of wild life and historical romance of the land are accepted material facts. More fascinating is the intangible charm and spell of this desert region where the haze of the distant mountain ranges meets the blue of the sky, and the desert impressive in its cloak of utter silence awaits the nature lover." Brown, with the assistance of U.S. senator Carl Hayden, sought to protect the Tucson Mountains, which were being threatened by homesteaders and mining interests. On April 11, 1929, the Pima County Board of Supervisors established Tucson Mountain Park, at that time the largest county park in the United States. (Courtesy of Wendall Brown.)

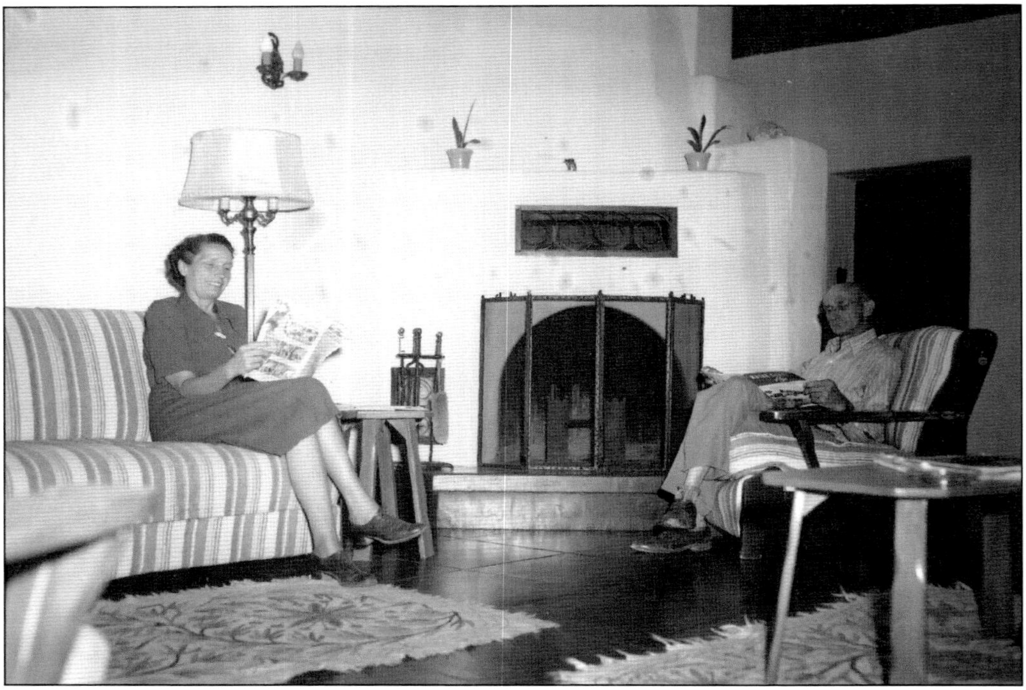

George and Alberta Zoellner were photographed seated in the living room of the Mountain House. Bill Carr described the building, "When I first saw this basically beautiful structure, it consisted of a large dining room and kitchen, a living room with a ceiling of exposed pine logs and saguaro ribs, two bedrooms, and a large patio. Across this clay-tiled patio from the dining room was another very large room, which was called a lounge. The ceilings in this main room and in the dining room were beams of large logs, roughly squared and adzed. It fit the landscape by being low, sprawling, and thus inconspicuous." A short distance up the sloping *bajada* from the Mountain House was a small house, a stable, and a corral. This photograph was taken of the main entrance to the stable. (Both, courtesy of Zoellner family.)

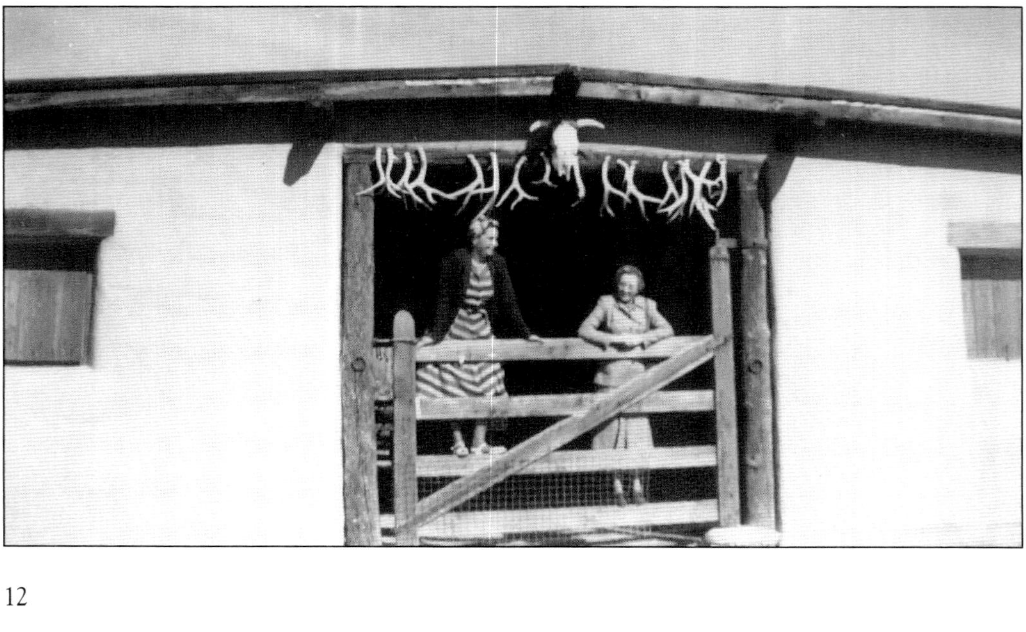

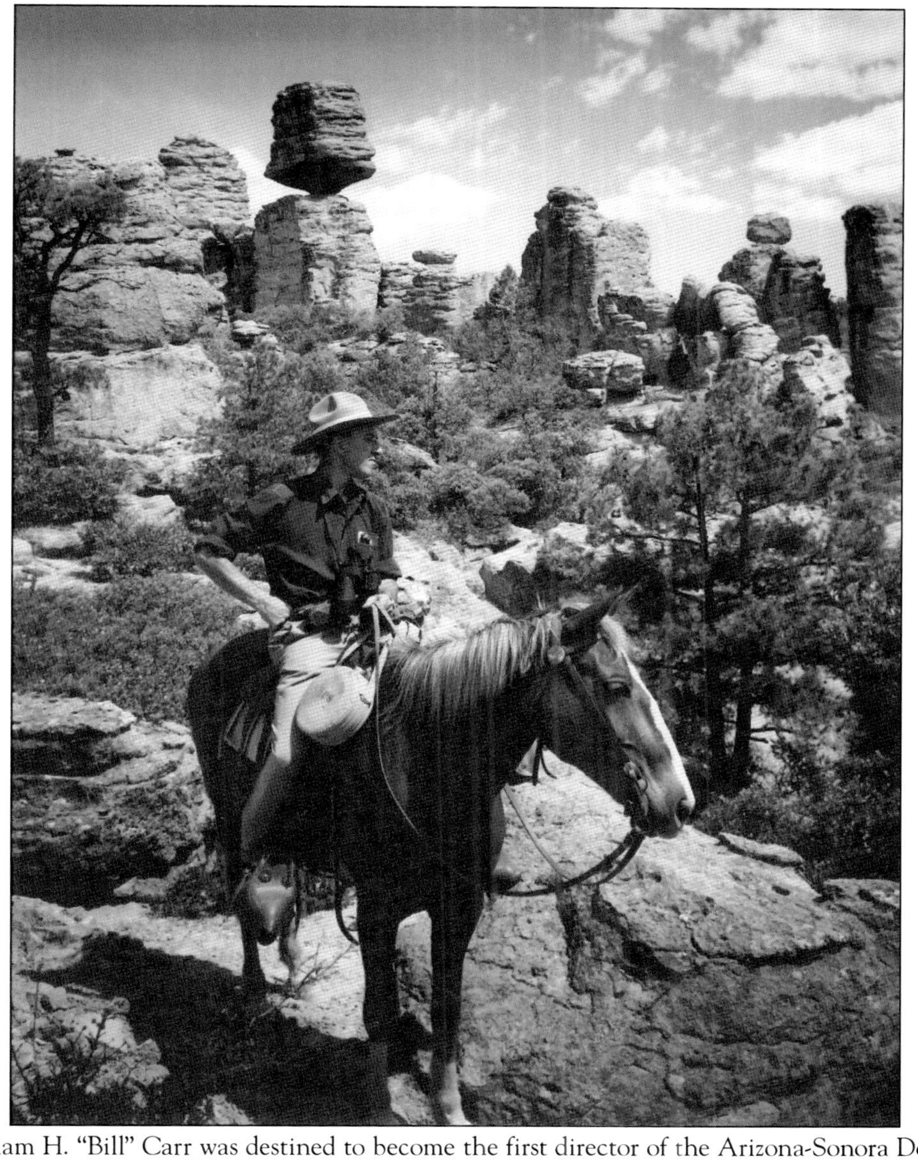

William H. "Bill" Carr was destined to become the first director of the Arizona-Sonora Desert Museum. He was unable to finish high school due to his father's death, but he stated that he gained his education at the American Museum of Natural History, where he first volunteered and then was hired in the education department. Under their auspices, he was assigned the task of establishing the Nature Trails and Trailside Museums at Bear Mountain in New York's Hudson River Valley. During his directorship there, he became an early, outstanding proponent of outdoor nature education. After 19 years, due to poor health and his doctor's advice, he relocated to the drier climate of Tucson, arriving in 1944 with "$400 and four suitcases." For a brief time after moving to Tucson, Carr was employed by the National Park Service. Here he is pictured at Chiricahua National Monument before beginning his campaign to found the Desert Museum. Carr dreamed of establishing a museum similar to Bear Mountain in the desert. With the strong support of Arthur N. Pack, the Pima County Parks Committee, and interested community members, Carr and Pack founded the Arizona-Sonora Desert Trailside Museum in 1952.

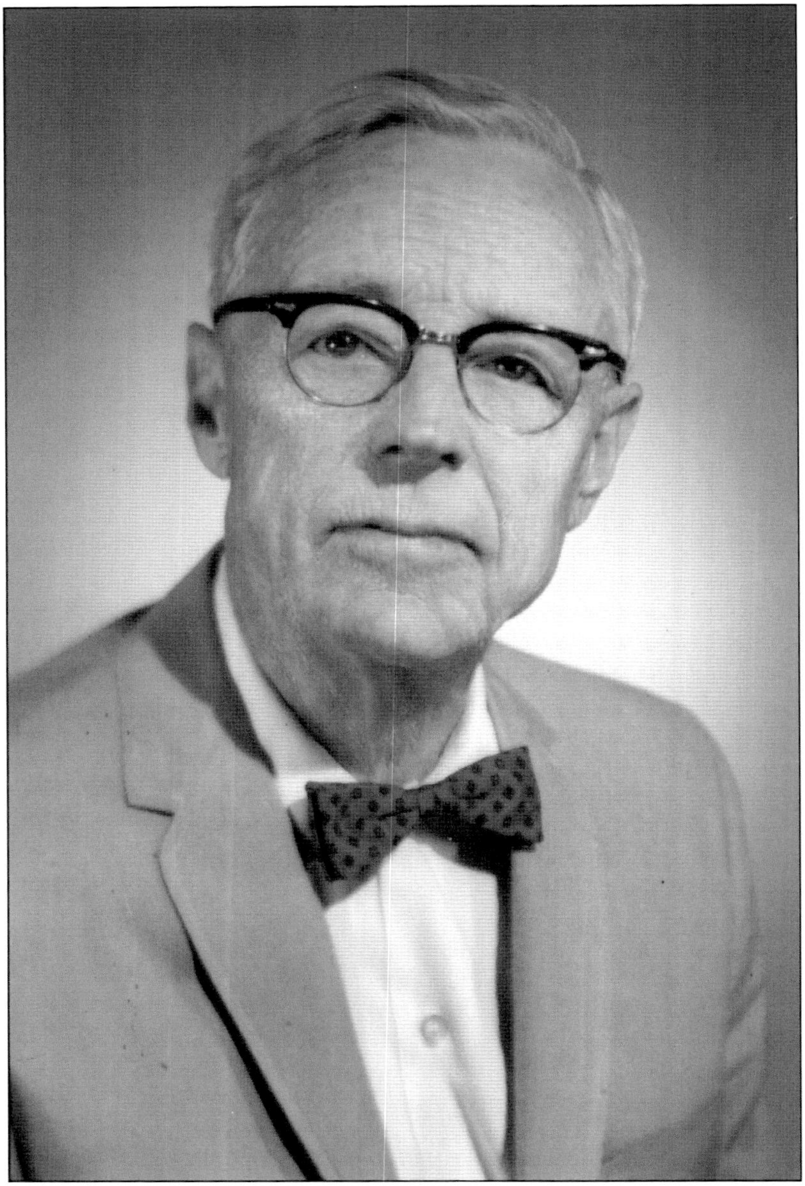

It was a serendipitous bit of fate when Arthur N. Pack and Bill Carr attended the same meeting of the Pima County Park Committee, as the members were debating the future of the Tucson Mountain House. Pack was a World War I veteran, then later editor of *Nature* magazine. This was a publication of the American Nature Association, which had been established by his father. The family had made its fortune in timber in the East and set up the Charles Lathrop Pack Forestry Foundation, of which Arthur was vice president. In the early 1930s, Pack established Ghost Ranch in New Mexico. The ranch hosted many visitors from the eastern United States and during World War II was a relaxing retreat for scientists working at nearby Los Alamos. After the war, Pack and his wife, Phoebe, relocated to Tucson. Following Carr and Pack's initial meeting, Arthur Pack provided funds, Bill Carr provided ideas and enthusiasm, and the two became cofounders of the Desert Museum.

Joe Carithers Jr., shown here with Bill Carr (left) on the museum patio, was superintendent of Tucson Mountain Park. He and his family lived near the Mountain House and had raised an orphaned deer and a javelina, or peccary, later released in the park. The two animals had bonded and wandered the area together, sometimes being fed by George Zoellner (below). Carr became acquainted with the unusual animal pair as the museum was being developed and wrote, "They often appeared from behind some shady corner and approached me very purposefully, the deer reaching into my pocket [for a hoped-for apple], the peccary nibbling at my trouser legs." Wistfully Carr recorded, "How wonderful it might have been if we could have exhibited all our animals in this same delightful manner, with them free to come and go as they pleased."

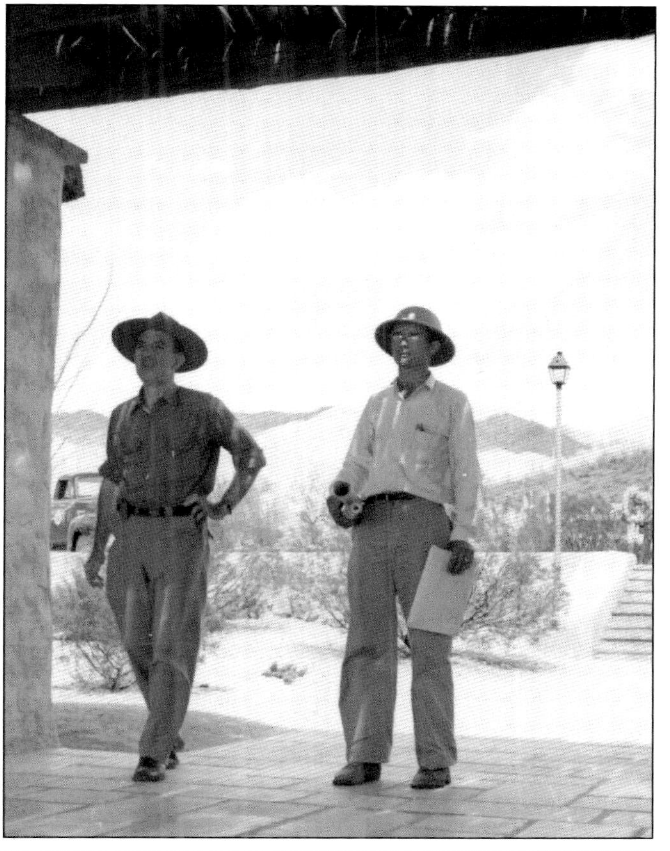

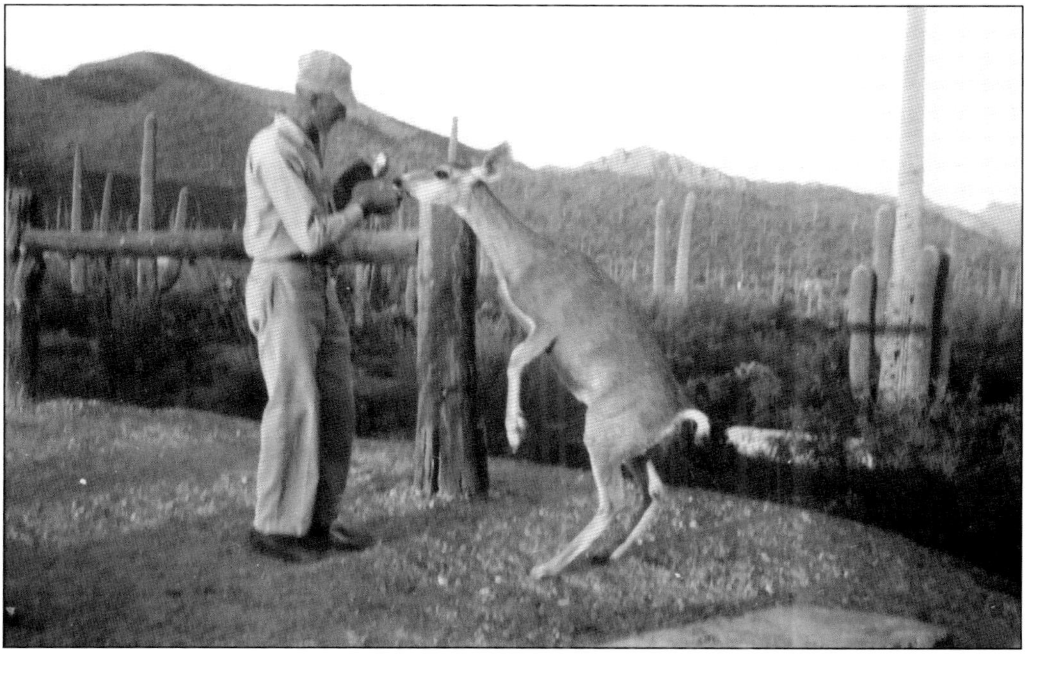

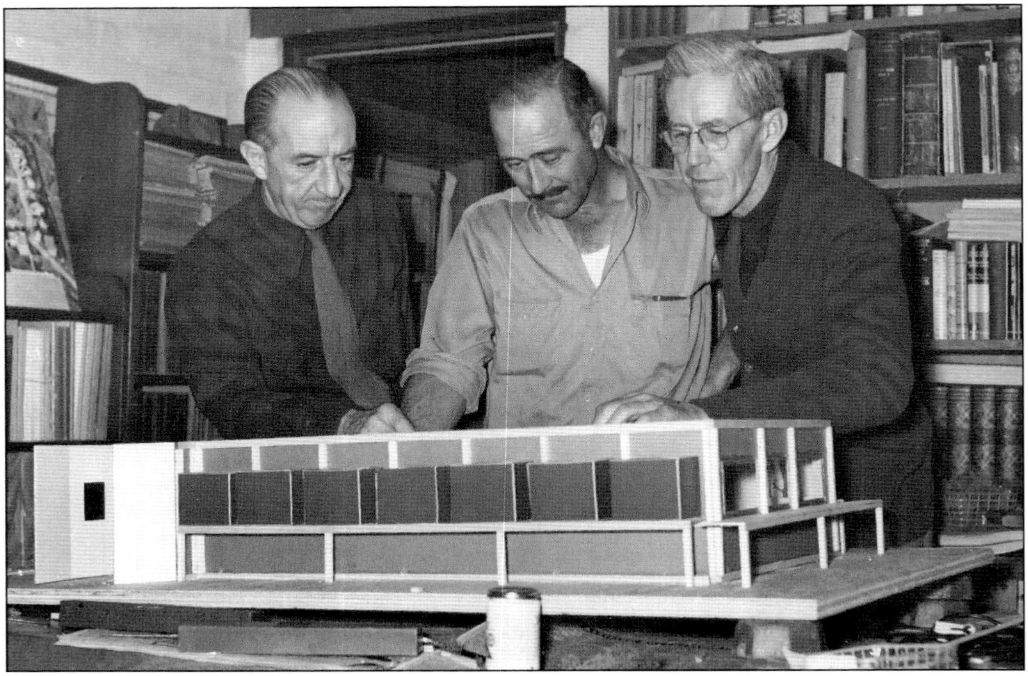

Once the right to use the Mountain House was obtained, Carr began a vigorous campaign to raise support and funds for the museum. Here Carr (left), George Olin (center), and Marvin Frost construct a model of the future small animal room and overall model of the museum grounds. Olin was a naturalist, photographer, and author who assisted in construction at the museum before its opening. Frost was a fine photographer and a friend of Carr's. Carr wrote that moving the models was "a huge and heavy affair. Lugging them around to the various presentations we gave was a trying chore, but it seemed to give our audiences the first real vision of what we had in mind." (Above, photograph by Marvin H. Frost.)

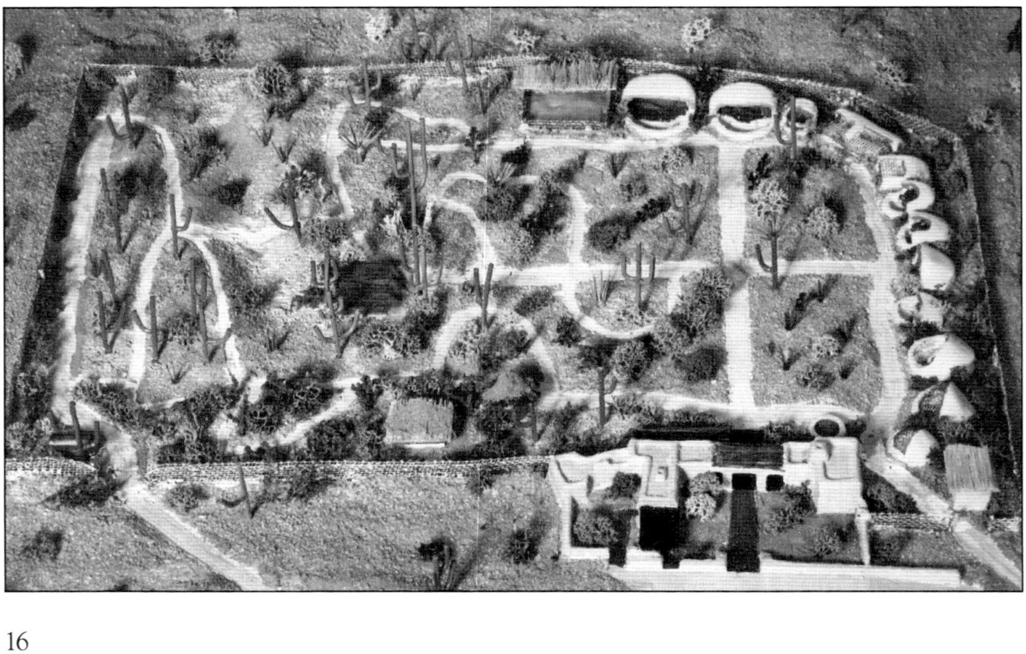

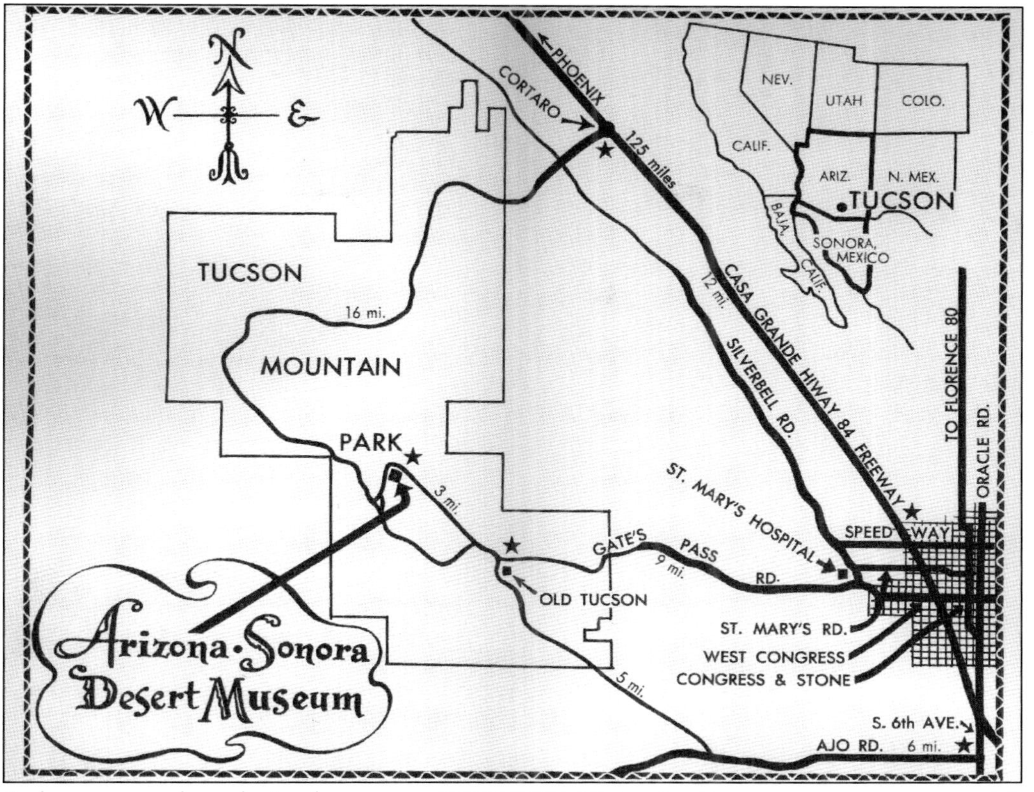

Early maps seen here depict the location of the Arizona-Sonora Desert Museum in relation to the city of Tucson (above) and also provide the viewer with some of the exhibits that existed during the first decade of the museum's operation (right).

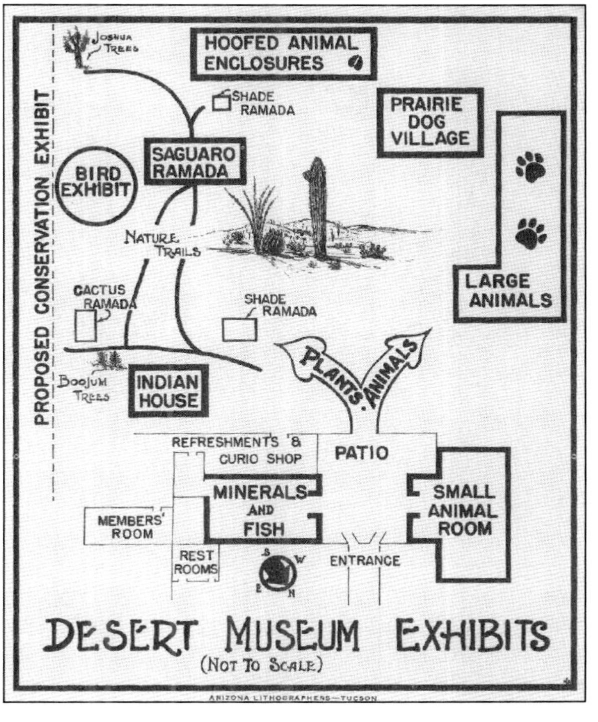

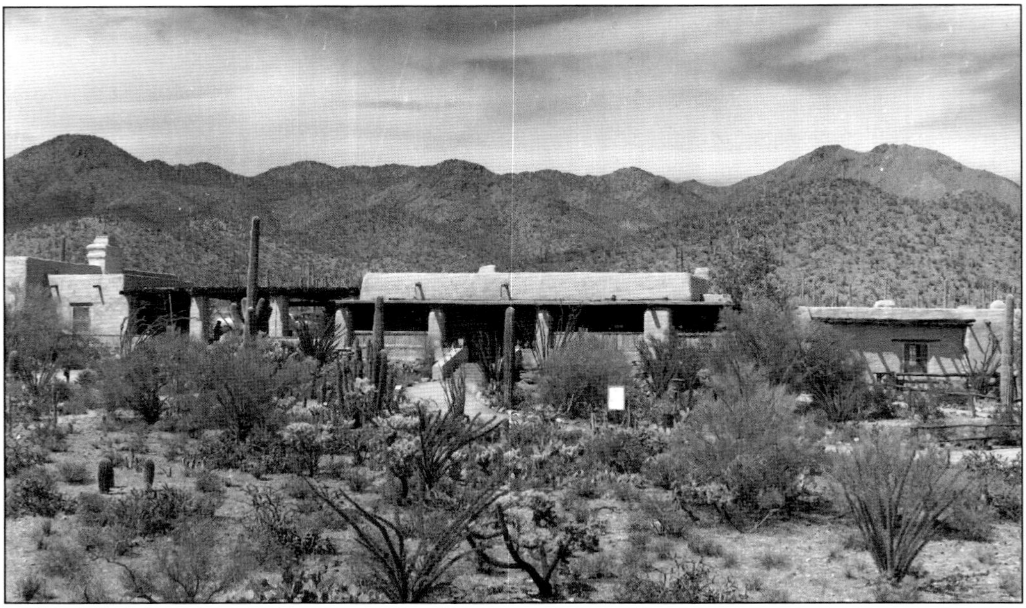

This photograph taken soon after the museum's establishment shows a portion of the small animal room on the left. The veranda connects it to the larger wing on the right. Along the latter runs an open porch. In early years this porch served as a gift shop, largely open to the elements. Now enclosed, it remains a gift shop adjoining the indoor-outdoor food concession, named Phoebe's in honor of Phoebe Pack.

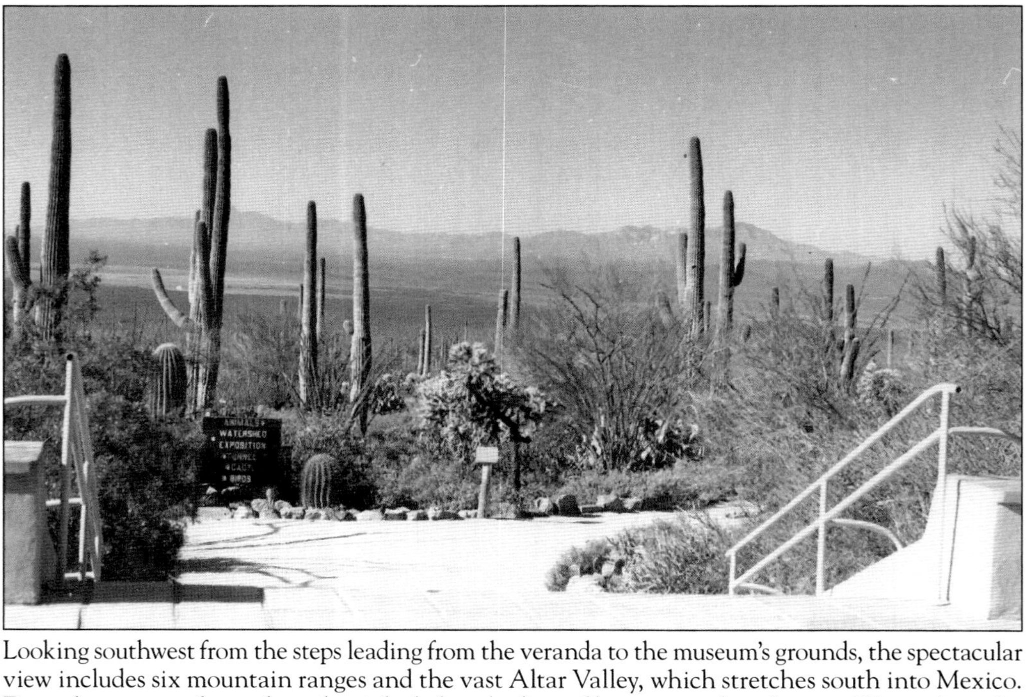

Looking southwest from the steps leading from the veranda to the museum's grounds, the spectacular view includes six mountain ranges and the vast Altar Valley, which stretches south into Mexico. From these steps, the path to the right led to the line of large animal enclosures. The path to the left led to the Tunnel, bird circulars, and Water Street.

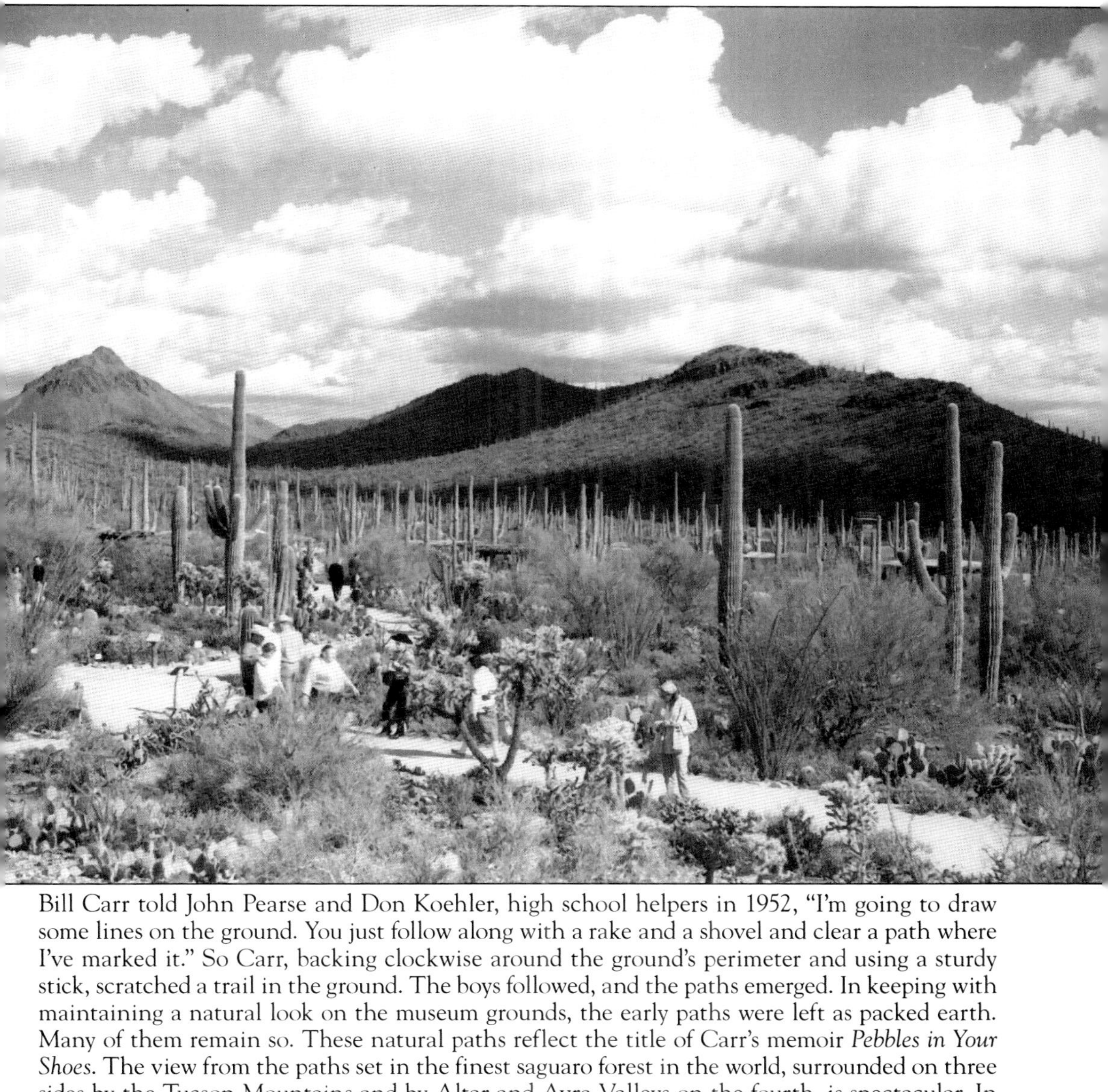

Bill Carr told John Pearse and Don Koehler, high school helpers in 1952, "I'm going to draw some lines on the ground. You just follow along with a rake and a shovel and clear a path where I've marked it." So Carr, backing clockwise around the ground's perimeter and using a sturdy stick, scratched a trail in the ground. The boys followed, and the paths emerged. In keeping with maintaining a natural look on the museum grounds, the early paths were left as packed earth. Many of them remain so. These natural paths reflect the title of Carr's memoir *Pebbles in Your Shoes*. The view from the paths set in the finest saguaro forest in the world, surrounded on three sides by the Tucson Mountains and by Altar and Avra Valleys on the fourth, is spectacular. In this photograph, summer monsoon clouds hover over Brown Mountain (right) named in honor of C. B. Brown, father of Tucson Mountain Park, and Golden Gate Mountain on the far left.

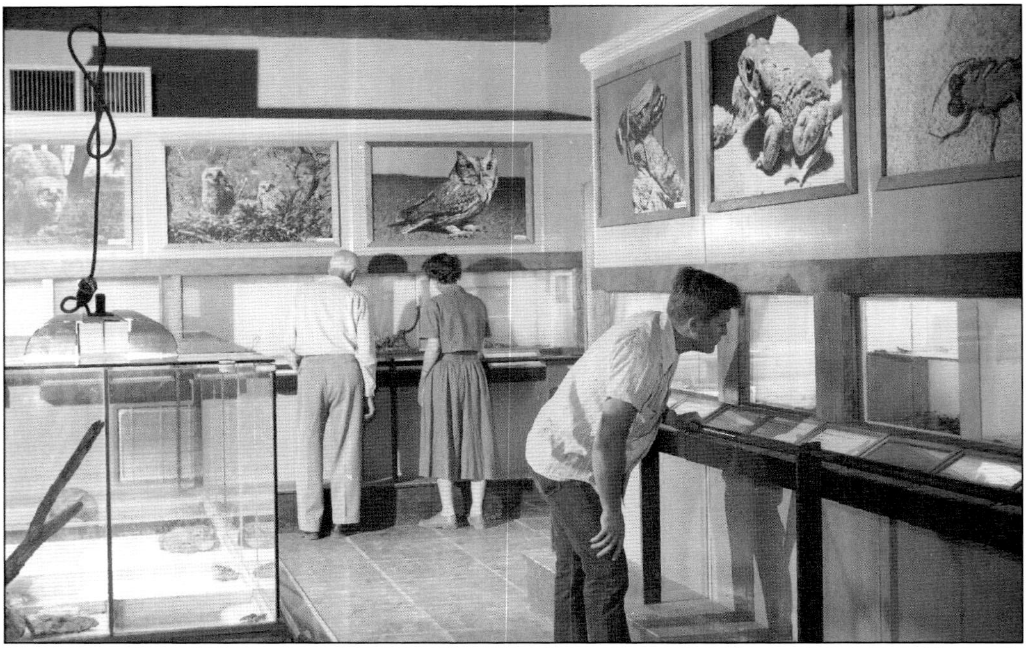

The original small animal room primarily featured reptiles in glass-fronted cages. Well-written labels provided detailed information, and large photographs taken by Marv Frost proved an added attraction.

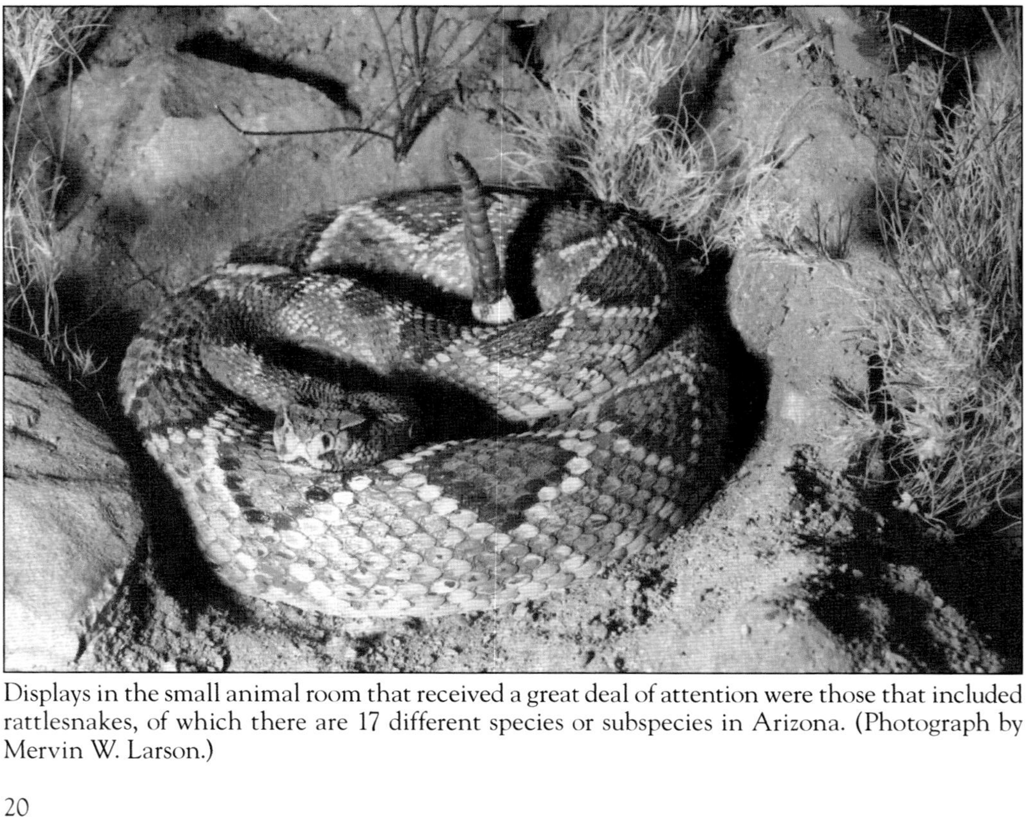

Displays in the small animal room that received a great deal of attention were those that included rattlesnakes, of which there are 17 different species or subspecies in Arizona. (Photograph by Mervin W. Larson.)

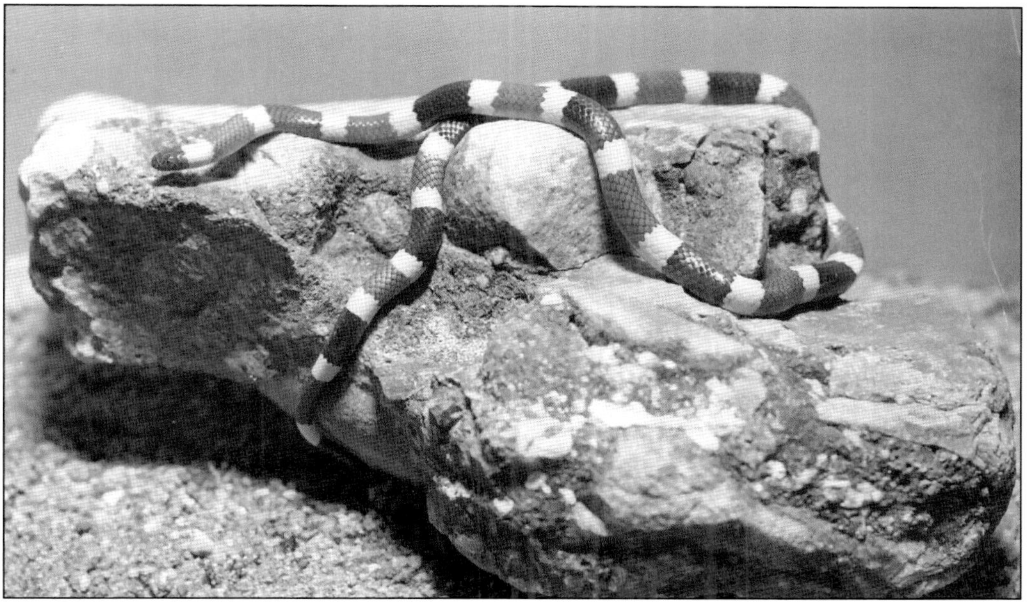

The Sonoran coral snake, although venomous, is much smaller than the better-known coral snake of the southeastern United States. Sonoran corals are only seldom seen. They have such small mouths it is difficult for them to inflict a bite on humans. (Photograph by Marvin H. Frost.)

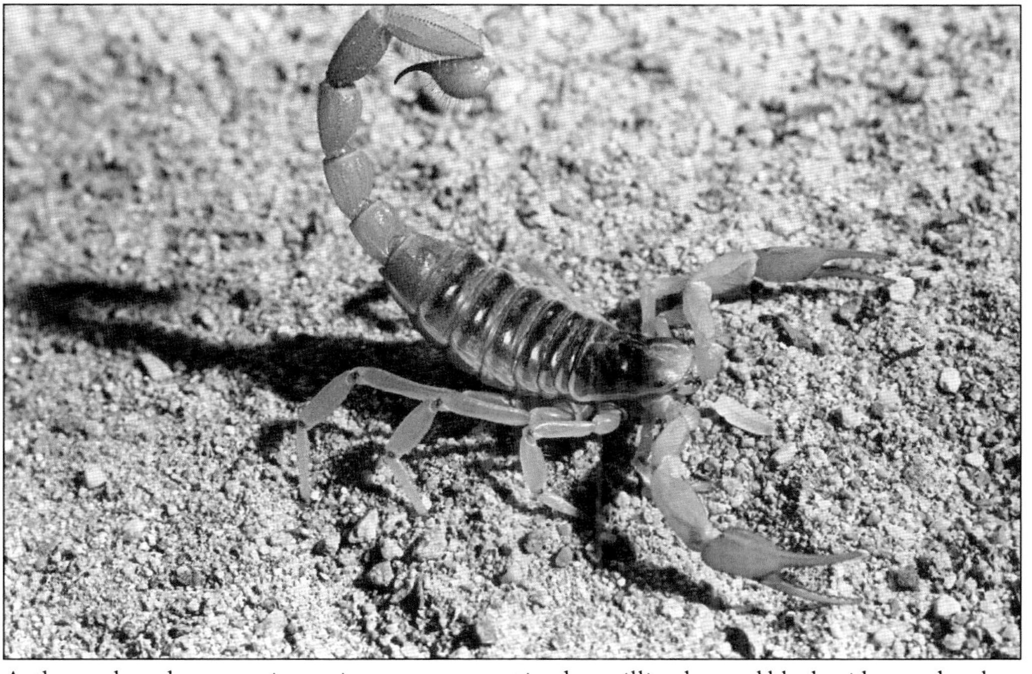

Arthropods such as scorpions, vinegarroons, centipedes, millipedes, and black widow and recluse spiders were displayed near the entrance to the small animal room. Visitors viewed three species of scorpions commonly found in the Sonoran Desert Upland region, in which the museum is located. Arizona is home to more than 30 species of scorpions. Pictured is the giant or desert hairy scorpion *Hadrurus arizonensis*, which is the largest species found in the United States.

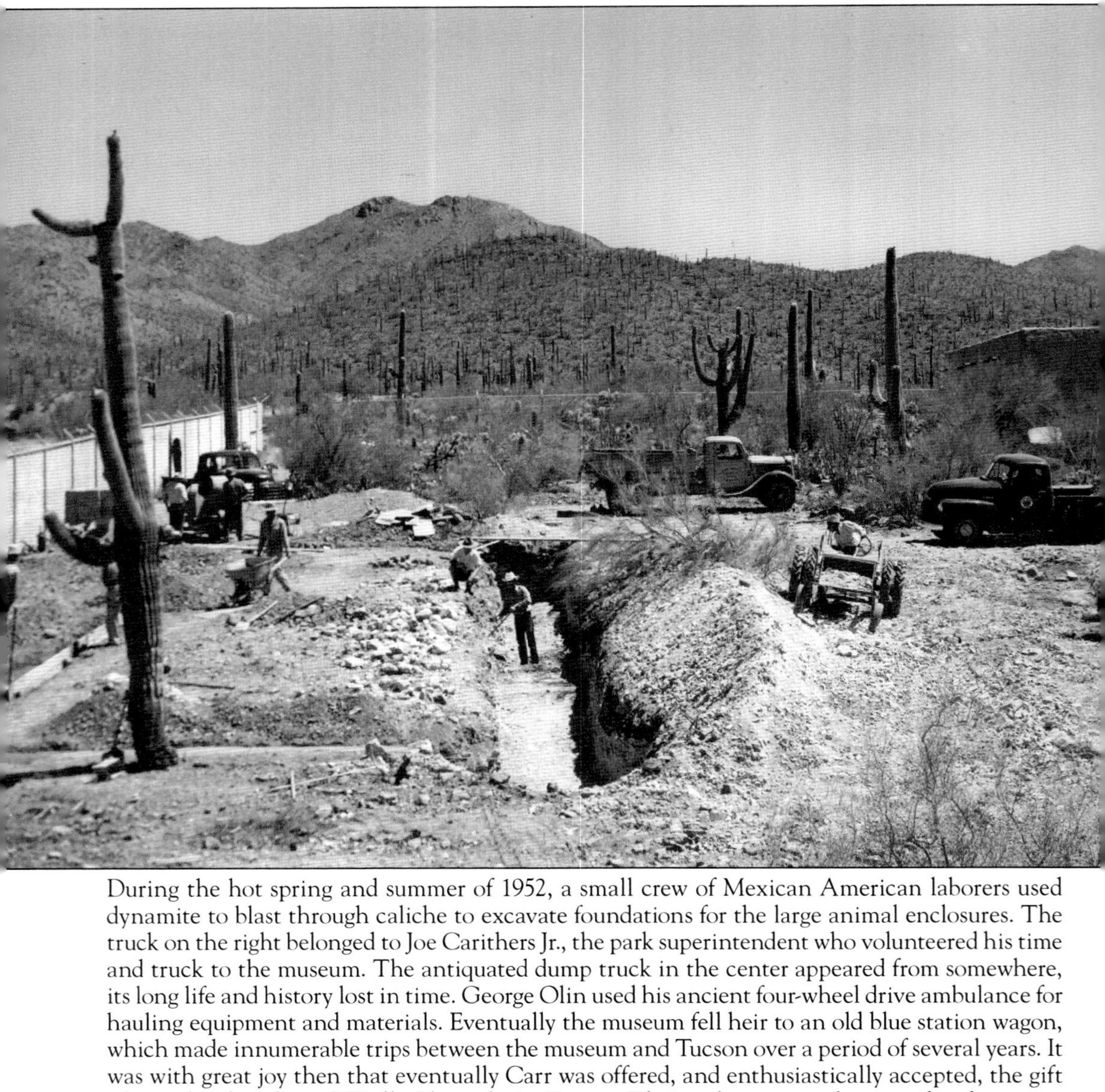

During the hot spring and summer of 1952, a small crew of Mexican American laborers used dynamite to blast through caliche to excavate foundations for the large animal enclosures. The truck on the right belonged to Joe Carithers Jr., the park superintendent who volunteered his time and truck to the museum. The antiquated dump truck in the center appeared from somewhere, its long life and history lost in time. George Olin used his ancient four-wheel drive ambulance for hauling equipment and materials. Eventually the museum fell heir to an old blue station wagon, which made innumerable trips between the museum and Tucson over a period of several years. It was with great joy then that eventually Carr was offered, and enthusiastically accepted, the gift of a used pickup from O'Rielly Chevrolet in Tucson. The truck was one of many gifts of a variety of types that came to the museum from the community.

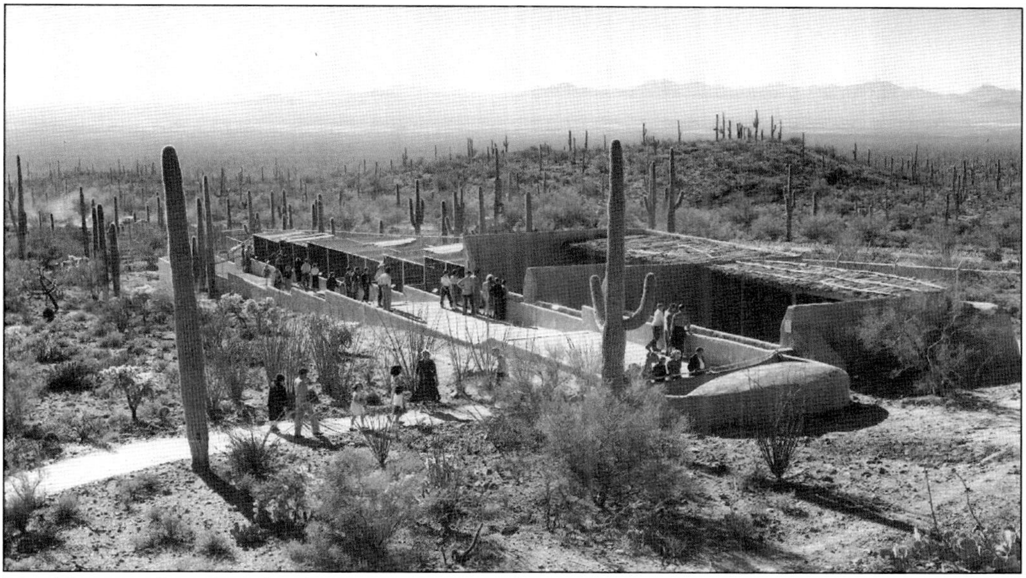

The blasting, shoveling, hauling, perspiring, and cement pouring resulted in enclosures for large mammals. The enclosures were built low to preserve the skyline, oriented to protect the animals from the western sun, and shaded for their comfort. Tortoises were housed under the low cement roof to the right in the upper photograph. That area eventually became the location of the bear enclosure.

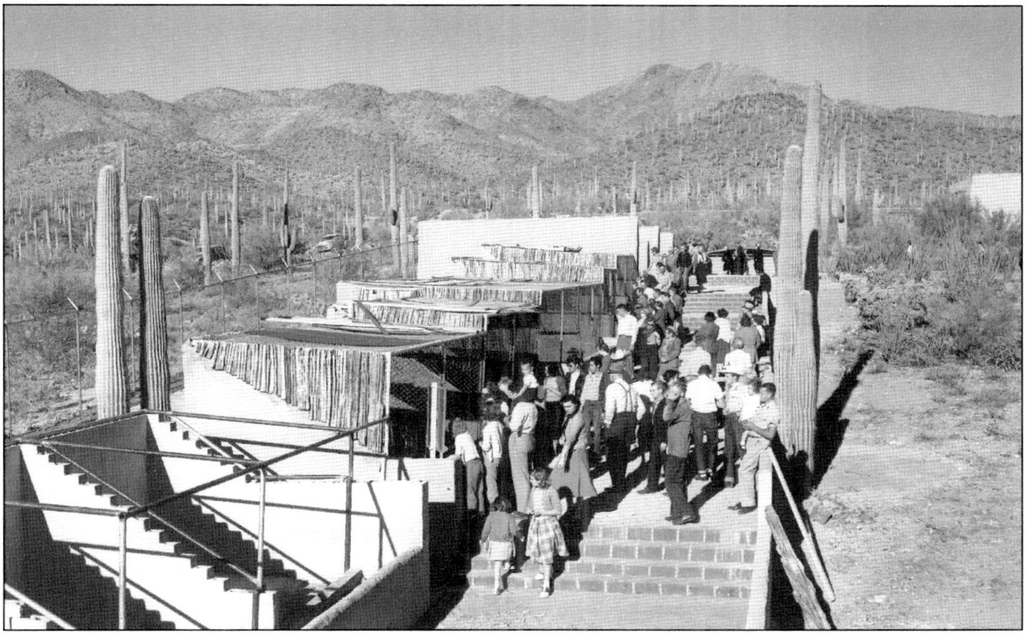

Crowds arrived on opening day and never stopped coming. Along what came to be known as the Large Animal Line, visitors viewed the mountain lions, bobcats, coyotes, javelinas, badgers, raccoons, and other mammals. The enclosures were relatively comfortable for the animals, easy to clean, and above average for that time, even though they had been built with limited funds. Through the years, they have been replaced with pleasing naturalistic habitats for the animals.

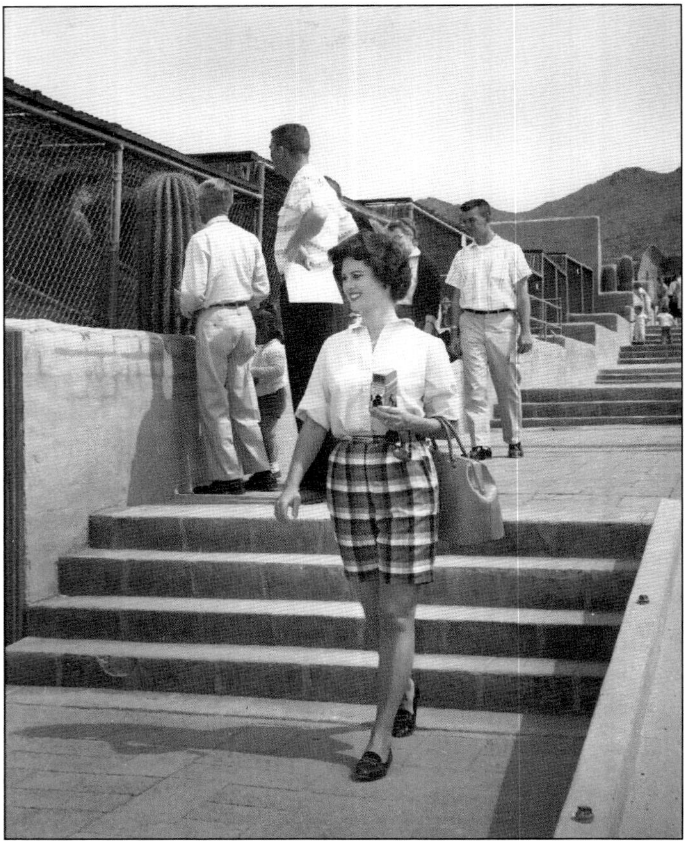

A stroll along the animal line was both pleasurable and educating. The Rolodex camera held by the visitor provides a good example of an early photographic instrument used to capture the moment during the early years at the museum.

A pair of javelinas was donated to the museum shortly before its opening after they had played a part in Disney's *The Living Desert* film. Just a few days before the museum's opening, while a class of university students was making a pre-opening visit to the museum, the female javelina gave birth before the watching audience. The javelina family was a hit for the opening day crowd. (Photograph by Lewis Wayne Walker.)

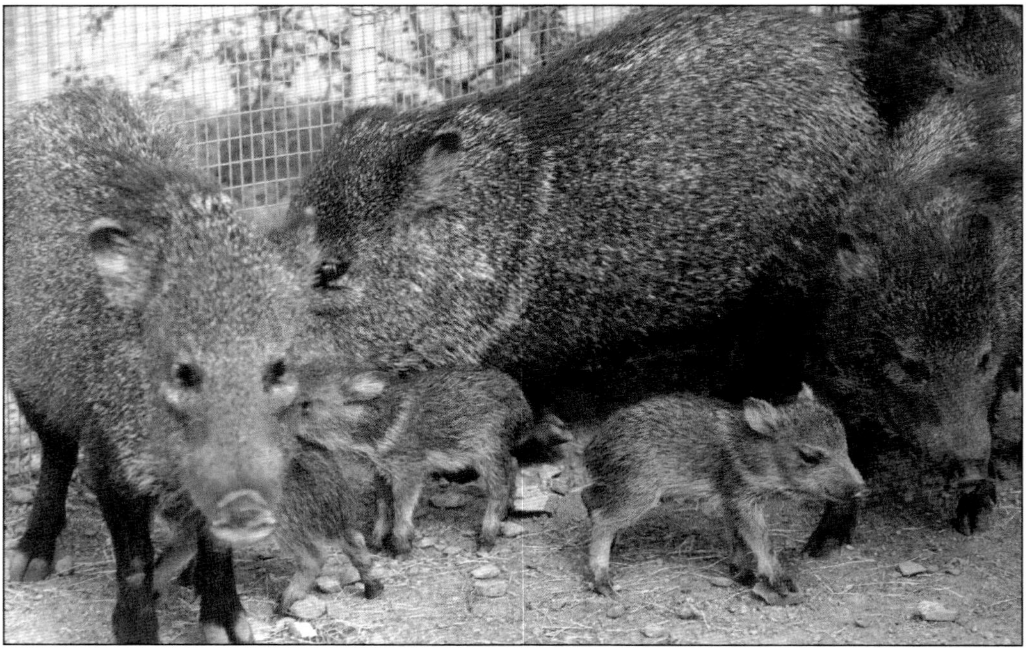

Also starring in Disney's movie was Tommy, an adult bobcat, who was filmed as he avoided pursuers by running to the top of a saguaro. At the museum, Tommy had a calm, reserved demeanor but often interacted with his caretakers. He liked to lick the Wildroot Cream Hair Oil from the top of one caretaker's head when that individual inadvertently walked under the elevated shelf on which Tommy often rested.

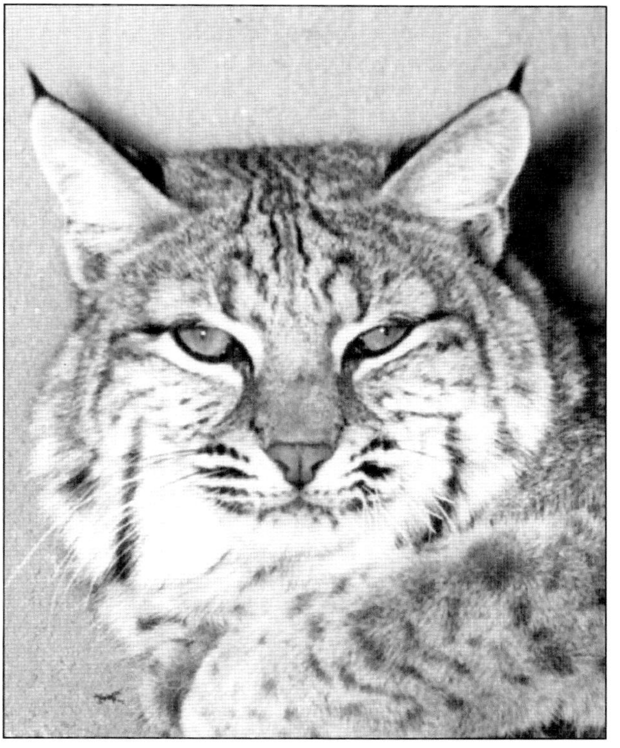

Family members of Arizona-Sonora Desert Museum employees were sometimes photographed with the animals on exhibit for purposes of publicity. The boy and the raccoon are a case in point.

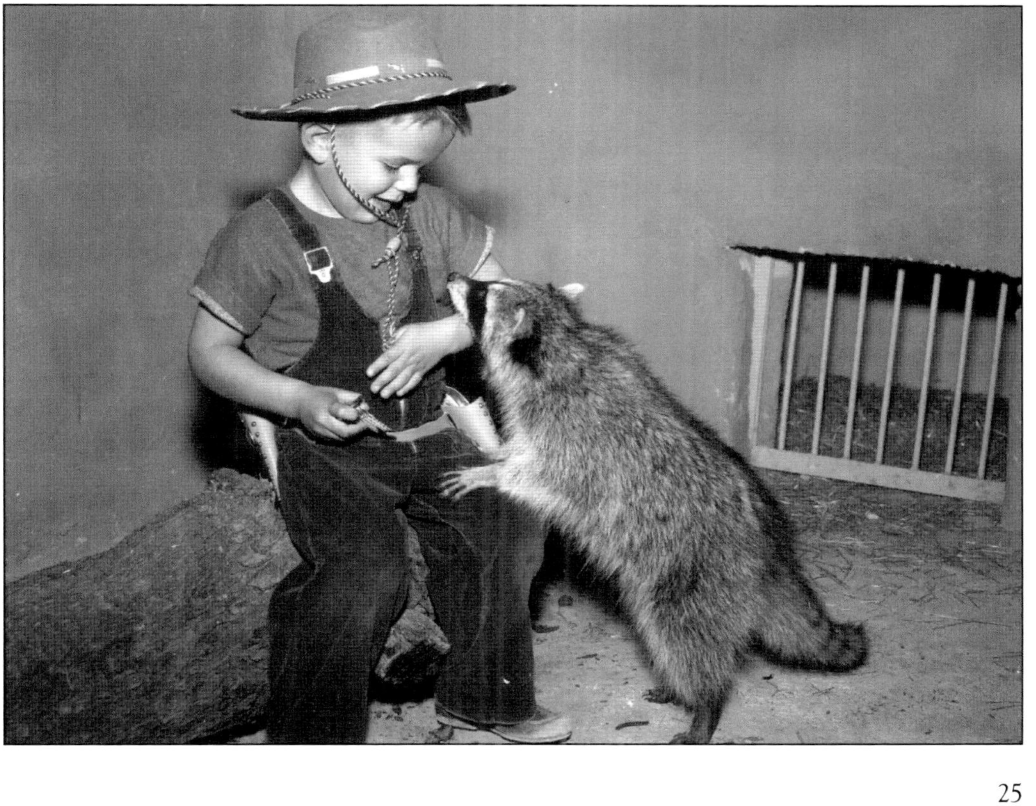

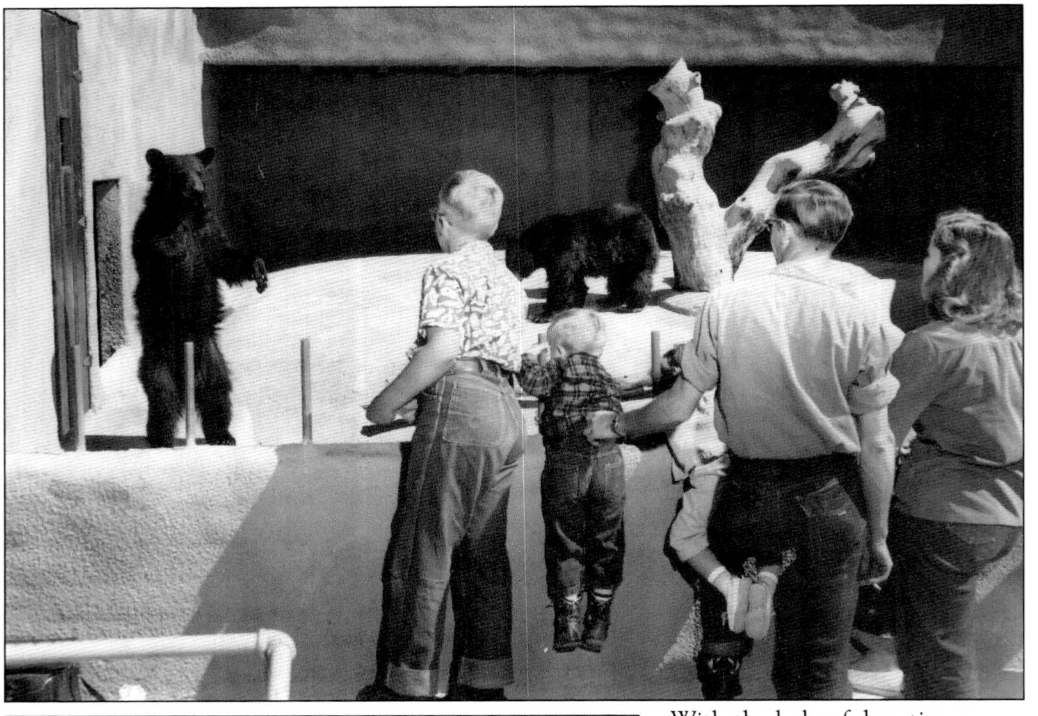

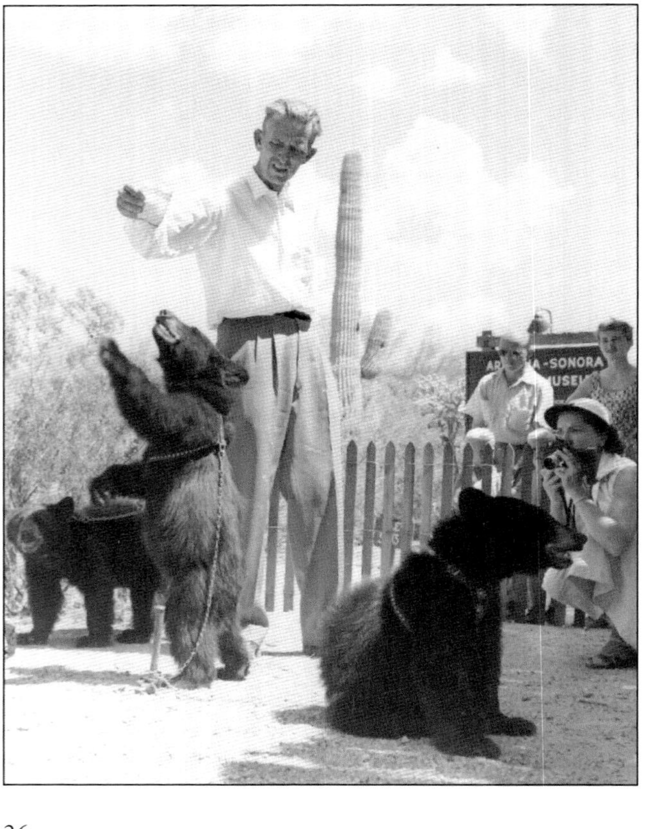

With the help of donations and the generous gifts of time and materials by the E. P. Hunziker Company and others, the bear exhibit was built soon after the museum's opening. Three young bears were popular early residents. Soon after the exhibit's completion, visitors could purchase bags of peanuts to feed to the bears. It entertained both species, until peanut shells clogged the drains. The practice was ended, and some serious replumbing was undertaken. A deep moat separated the viewers from the acrobatic bears. The stump was sturdy and almost bear-proof. Years later, the bears were released into a new natural habitat enclosure with transplanted living trees. Feeding the bears (left) is Lew Walker, who was the associate director of the museum for many years until his retirement in 1970.

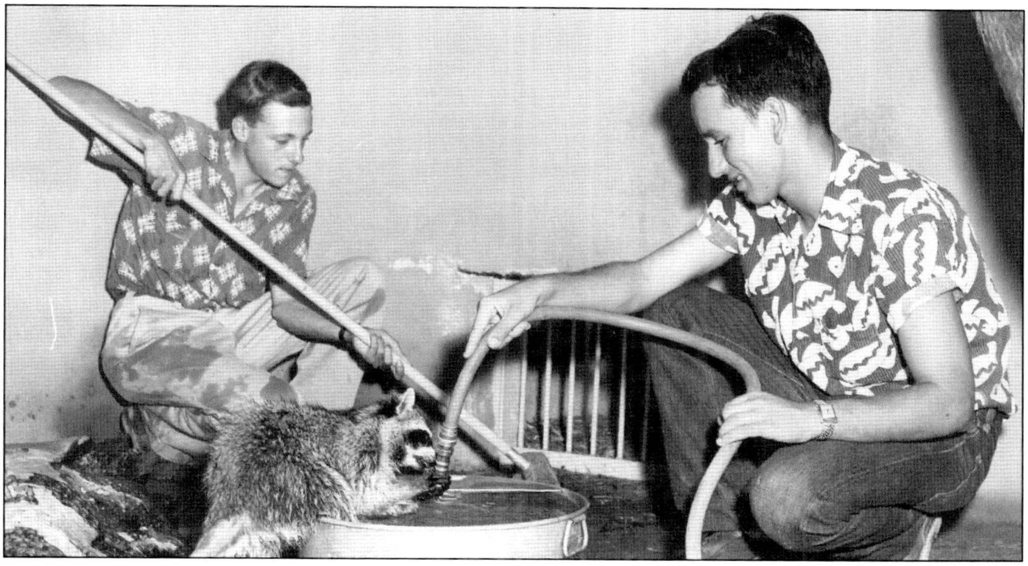

The cement animal enclosures were easily cleaned. Here a soggy raccoon supervises while Merritt Keasey wields the broom and John Rodi the hose. The enclosures were definitely not picturesque, sometimes suffering from peeling plaster or similar defects, as does this coyote's enclosure, and were fronted with ugly chain link fences. There was nothing that gave an indication of the natural environments in which the animals normally lived. Cement floors, plastered walls, and elevated branches or boards for the animals' resting places were far from ideal. In future years, these enclosures would be replaced by more comfortable quarters displaying animals in naturalistic habitats, a Desert Museum innovation that set new standards for animal exhibits in zoos and aquariums in the United States and many other parts of the world.

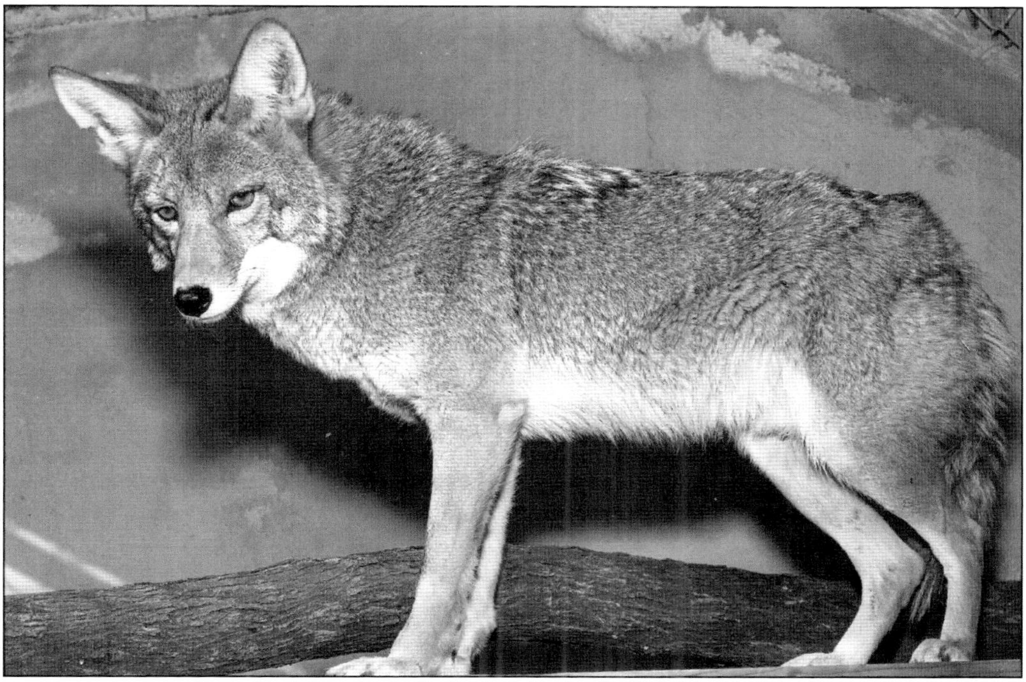

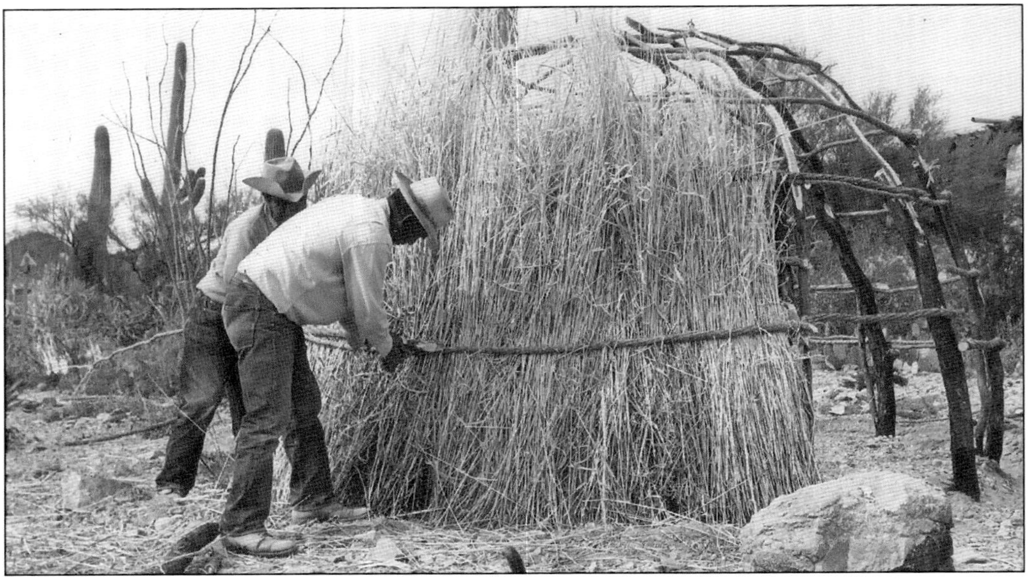

The Papago tribe (now the Tohono O'odham) lives on a large reservation near the museum. In order to provide visitors with information regarding the way these people ancestrally lived in and with the desert, Carr contracted with Frank Lopez (Ban-i-qIash or Peeping Coyote) to build a *shaish-ki*, a type of temporary home formerly used by the Papago. Mesquite branches were bent to form a conical framework, to which bundles of grass were bound with ocotillo branches. Outside the finished house, the men built a shade ramada (below). Pots, metates, manos, and other household items were arranged outside. The tall pole pictured in the center was made of portions of the saguaro skeleton and used to pull the fruit down from the saguaros. A mound of soil in the right foreground was being molded by the Lopez women into an outdoor oven.

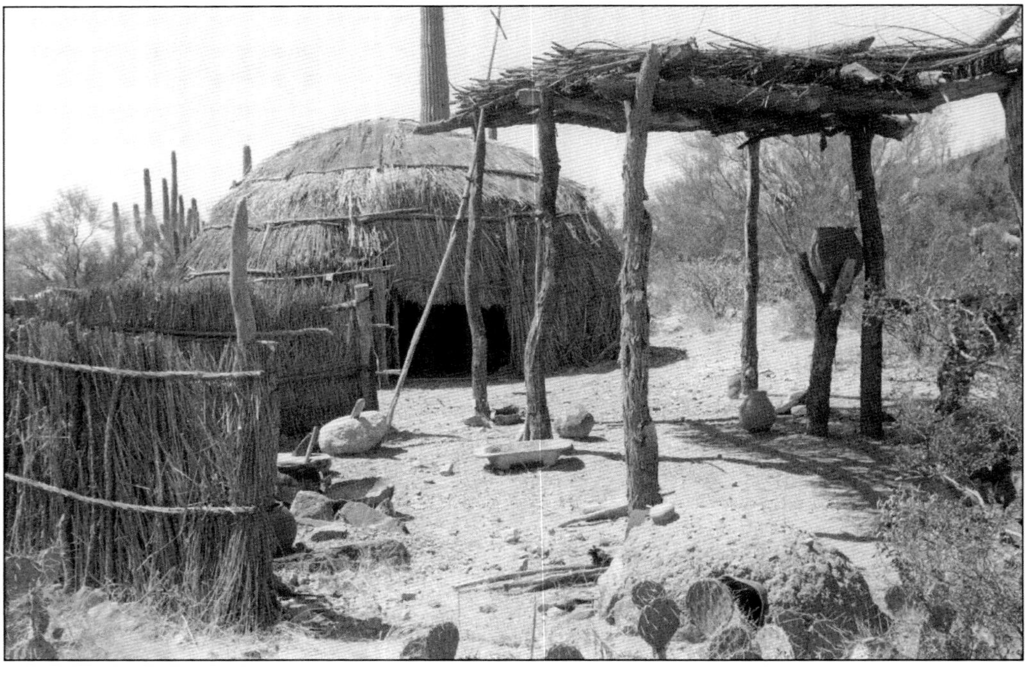

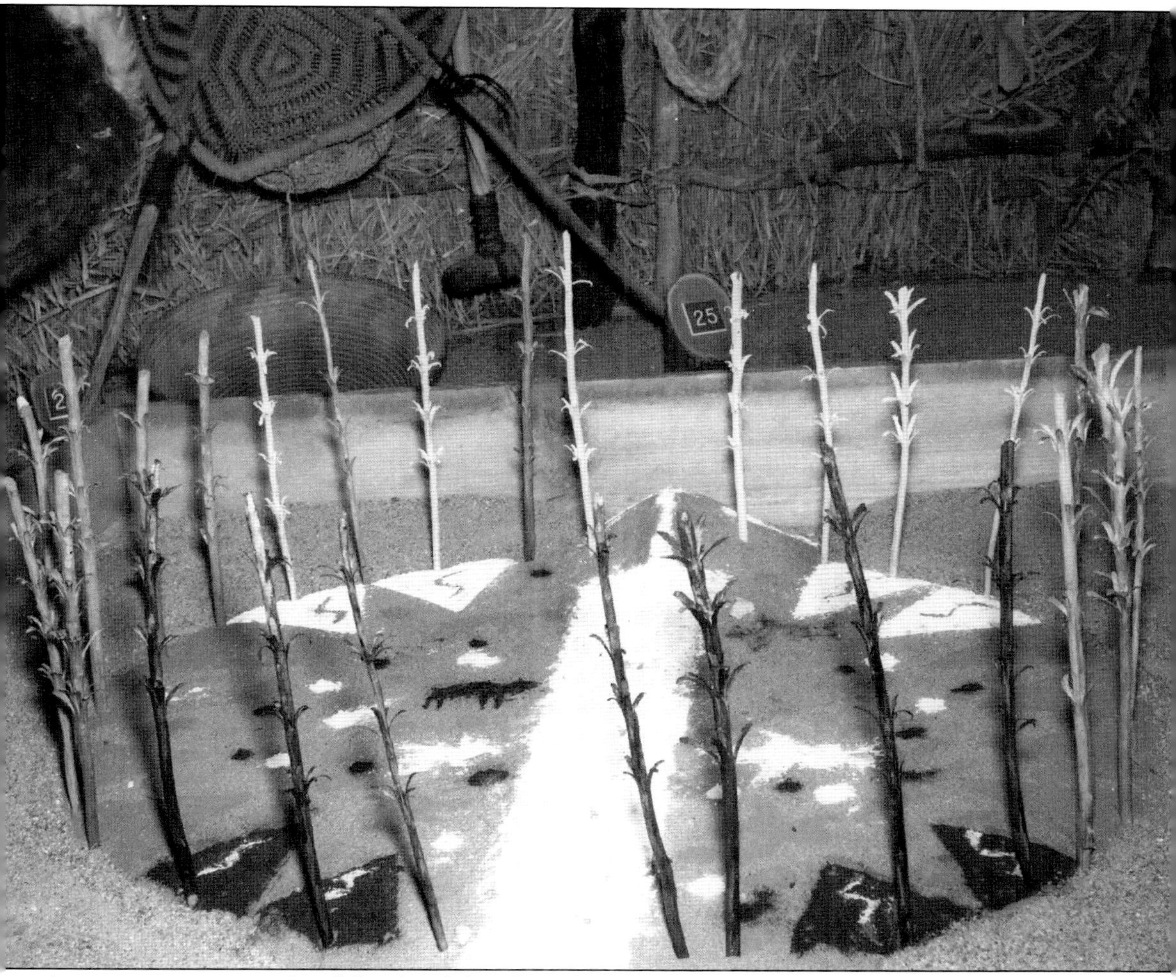

An opening in one side of the shaish-ki allowed visitors to view the interior, where various cultural items were placed. In the center of the floor, Lopez created a special sand painting. Sand paintings are considered of great importance to various Native American groups, and their use is often very private and tied to spiritual beliefs. Lopez told Carr that he would leave out the part that made the painting holy. Carr described the process used in producing the painting, "First Lopez made a low, level mound of packed earth, scratching designs in its top with a sharp stick. These grooves were later filled with colored material, for the most part sand that had been dyed. There were also colored clays and pigments made from plants, all of flat Indian colors: white, green, red, black and blue." Ordinarily Tohono O'odham sand paintings are destroyed at the end of the day in which they are built, but the museum obtained permission to leave this one permanently on display.

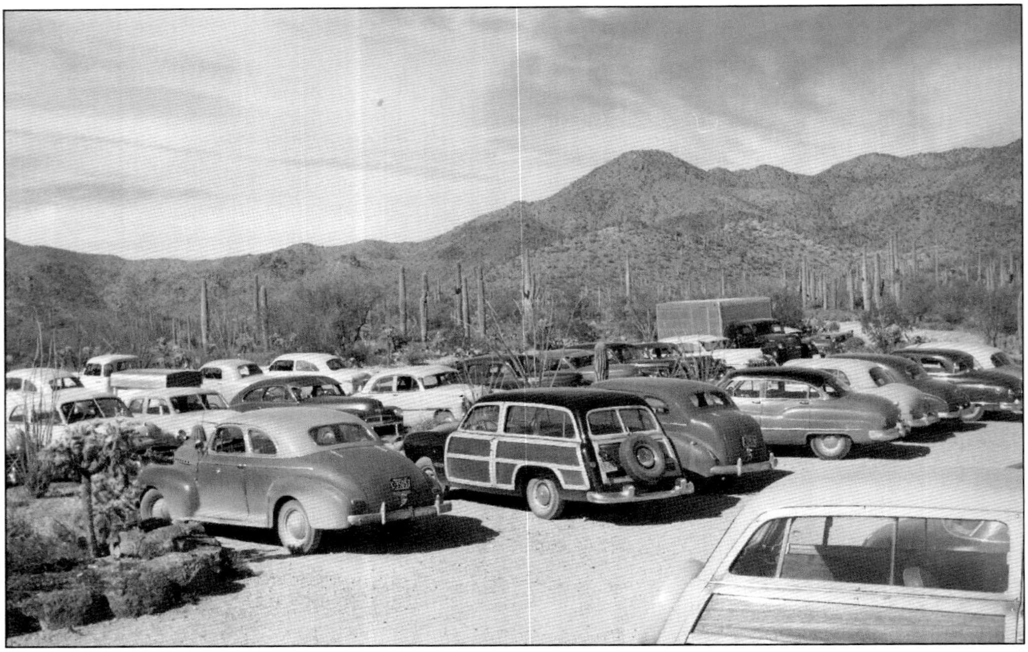

The museum's first parking lot was the one constructed for the original Mountain House and space was limited. On opening day it was soon filled, and cars were parked along the road for a half-mile in each direction. By February 1956, when the picture below was taken, a large second lot had been added. On Sundays, the museum's busiest days, dust often rolled across the lots, until they were finally paved. So numerous were vehicles on Sundays that at one point a group of Boy Scouts volunteered to direct parking in order to squeeze the most vehicles into the least space. The neat compact parking in the later picture indicates that Boy Scouts may have been at work on that particular Sunday.

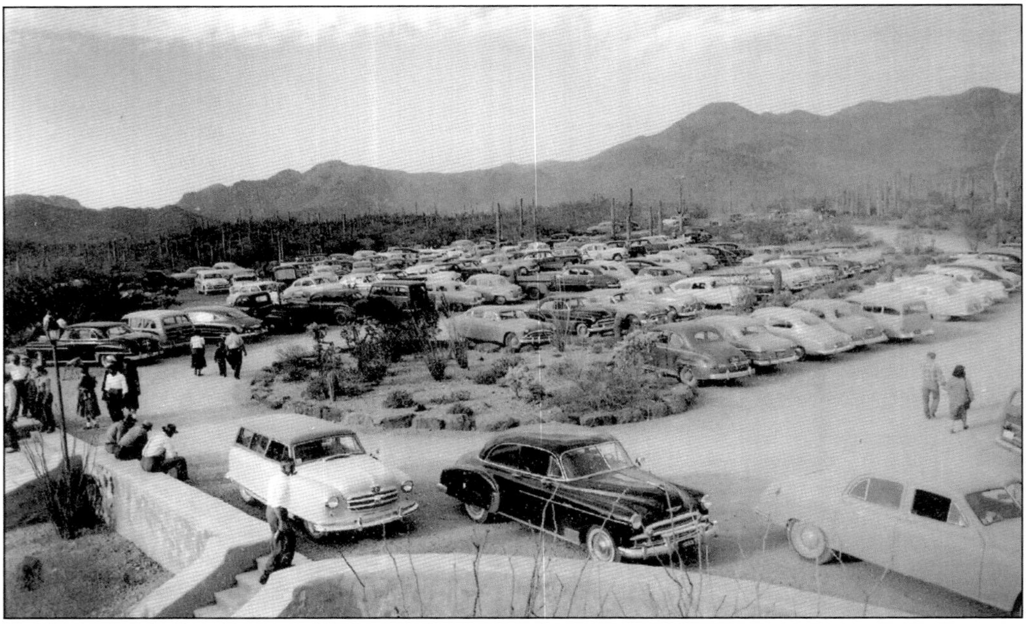

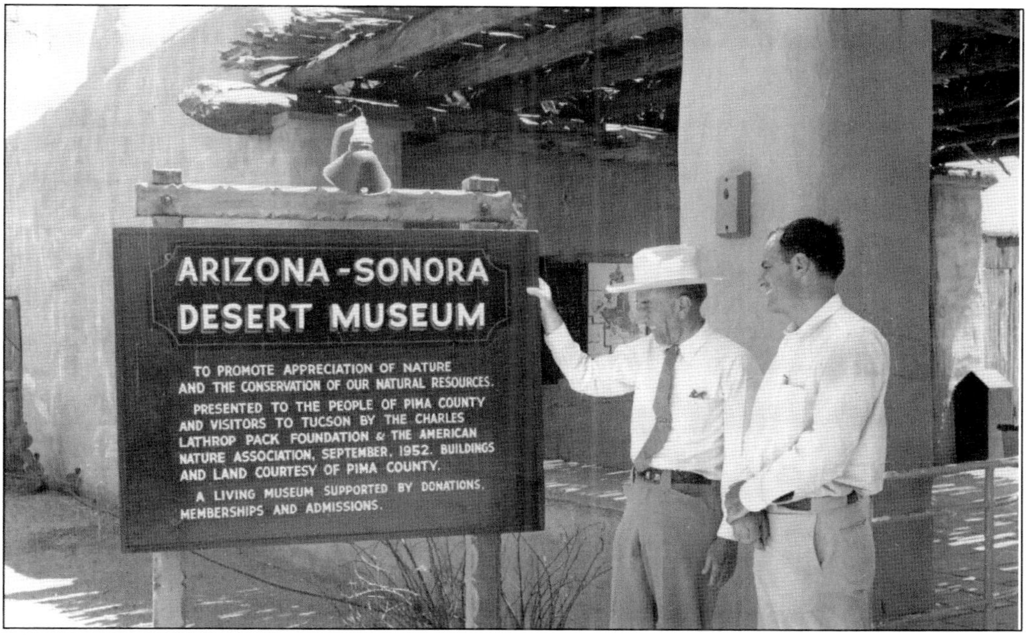

The first two directors of the museum, William H. Carr (left) and William H. Woodin were photographed at an early date standing at the edge of the veranda with the small animal room to their right. The succinct messages on the sign regarding goals, origins, and a nudge toward financial support said it all.

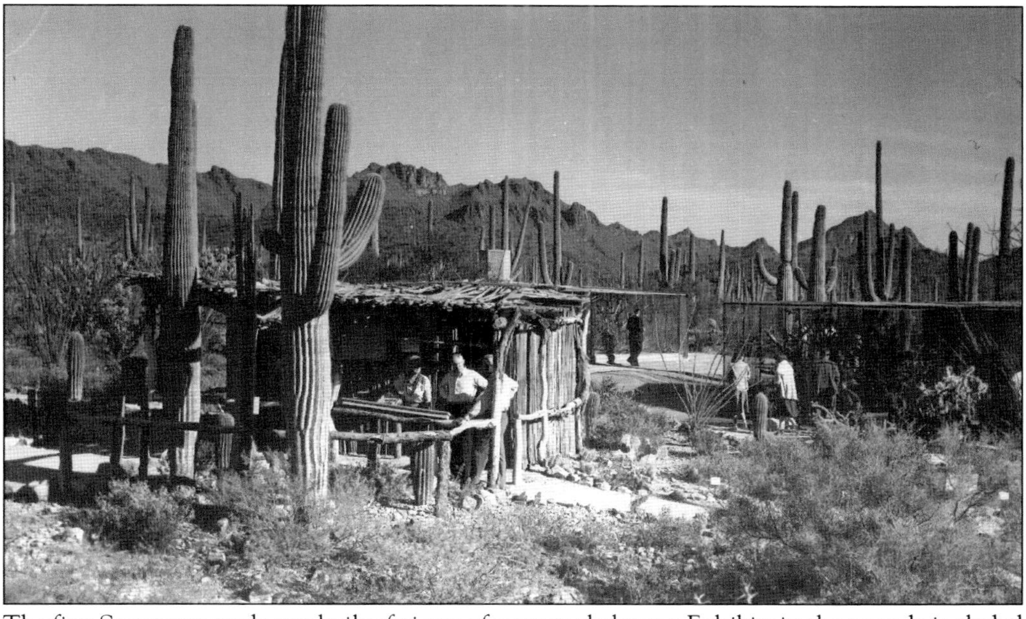

The first Saguaro ramada was built of pieces of saguaro skeletons. Exhibits in the ramada included tiny saguaro seeds, a section of a saguaro trunk showing the deep hole made by a Gila woodpecker, saguaro "boots," artificial saguaro flowers, and articles used by the Tohono O'odham in gathering saguaro fruit. Funds for this ramada and the more formal one that succeeded it were primarily donated by local garden clubs. (Photograph by Wyatt Davis.)

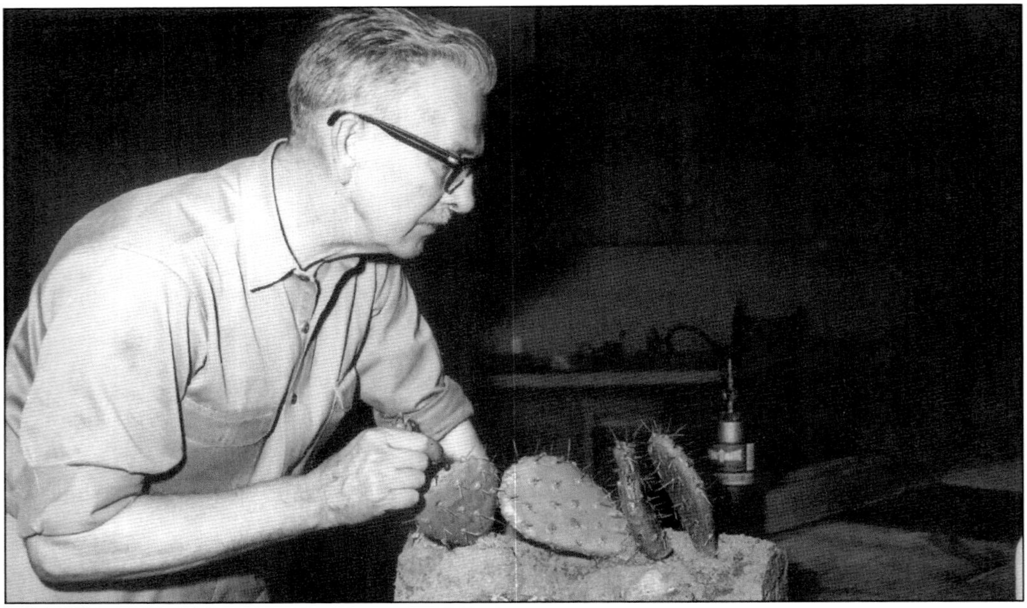

Lewis Wayne Walker joined the museum staff in 1953. Lew was an ornithologist, a photographer, a prolific writer on nature subjects, and highly adept at building exhibits, including this portion of one with prickly pear pads and a rattlesnake. Lew had known Carr when they were teenagers in New York, had worked for the San Diego Natural History Museum, the San Diego Zoo, and was in the service during World War II when he wrote a book about survival skills and taught them to American servicemen in the Pacific. (Photograph by Mervin W. Larson.)

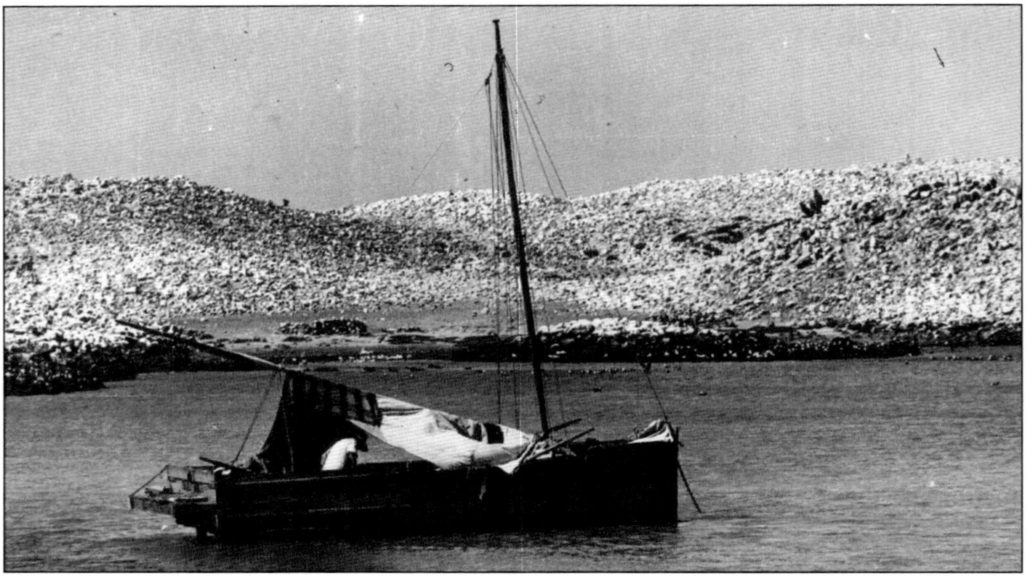

As early as the mid-1930s, Walker was exploring the Sea of Cortez. During his tenure at the museum, he was a moving force in promoting the Mexican government to declare Isla Rasa in the Gulf a refuge. He built this boat in the late 1940s to sail to Rasa. The vessel was supposed to be called the *Sacred Cow*, but the painter from San Felipe transposed the letters and painted on the name *Scared Cow*.

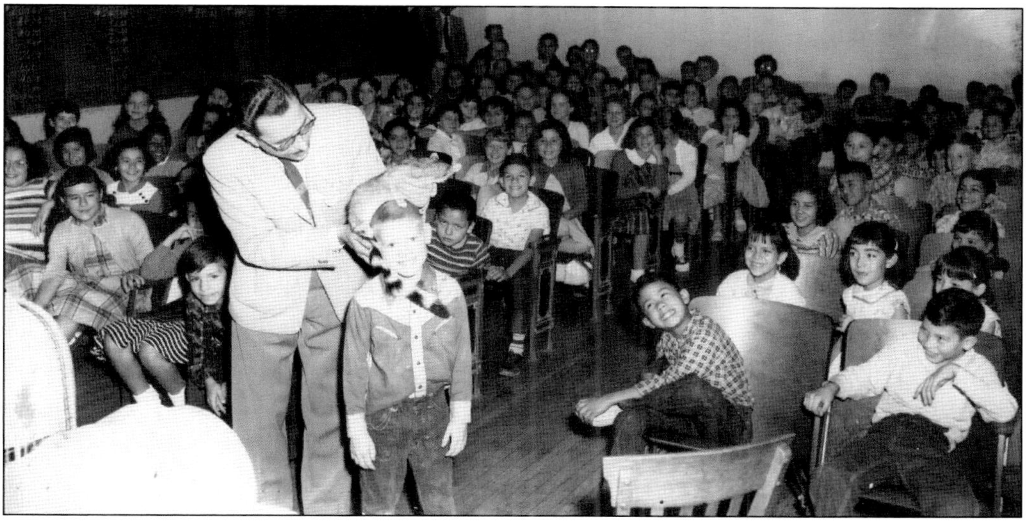

Hal Gras is providing an animal interpretation for which he was renowned among the local Tucson schools. This early presentation depicts the excitement and enthusiasm that children had for the novel concept of interpreting live animals in the classroom. Below is Hal Gras holding a javelina beside his chief mode of transportation, the Desert Ark. Over the years, a local Civitan Club donated a total of six such vehicles to be used as Arks.

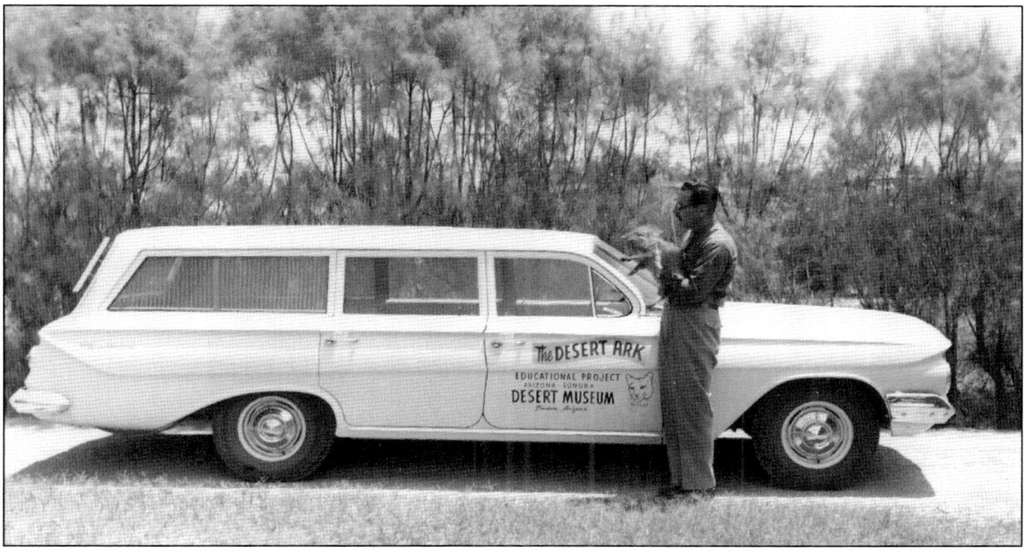

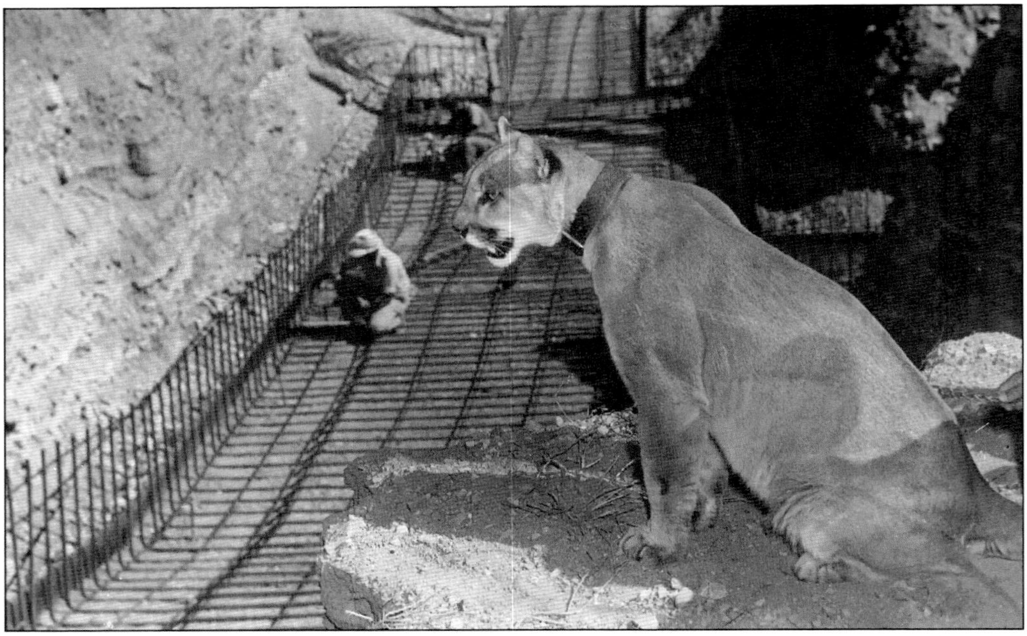

In 1956, Carr, acting as a consultant to the museum, obtained special funds from the Pack Foundation to adapt a new museum exhibit technique on a large scale by building an underground tunnel in which animals that had escaped from the heat on the desert floor could be viewed in their insulated underground retreats. Above, George L. Mountainlion II observes the beginning construction of the tunnel. The Tunnel, seen below, was 160 feet long, the walls 12 feet apart, and the ceiling 9 feet high. The project used 160 cubic yards of concrete, 8.5 tons of reinforcing steel and, inside, 22,000 pounds of casting plaster. Carr commented, "Quite an undertaking at that time for a once-lonely museum in the desert." Aboveground, the tunnel was covered by a mound of dirt and rock planted with cactus. (Both photographs by Hiram L. Parent.)

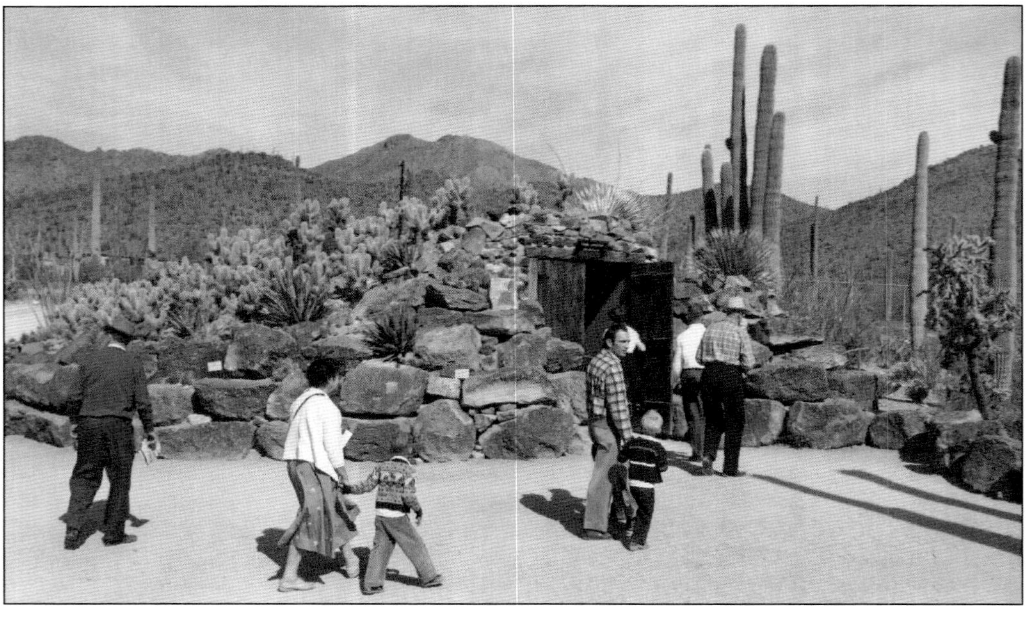

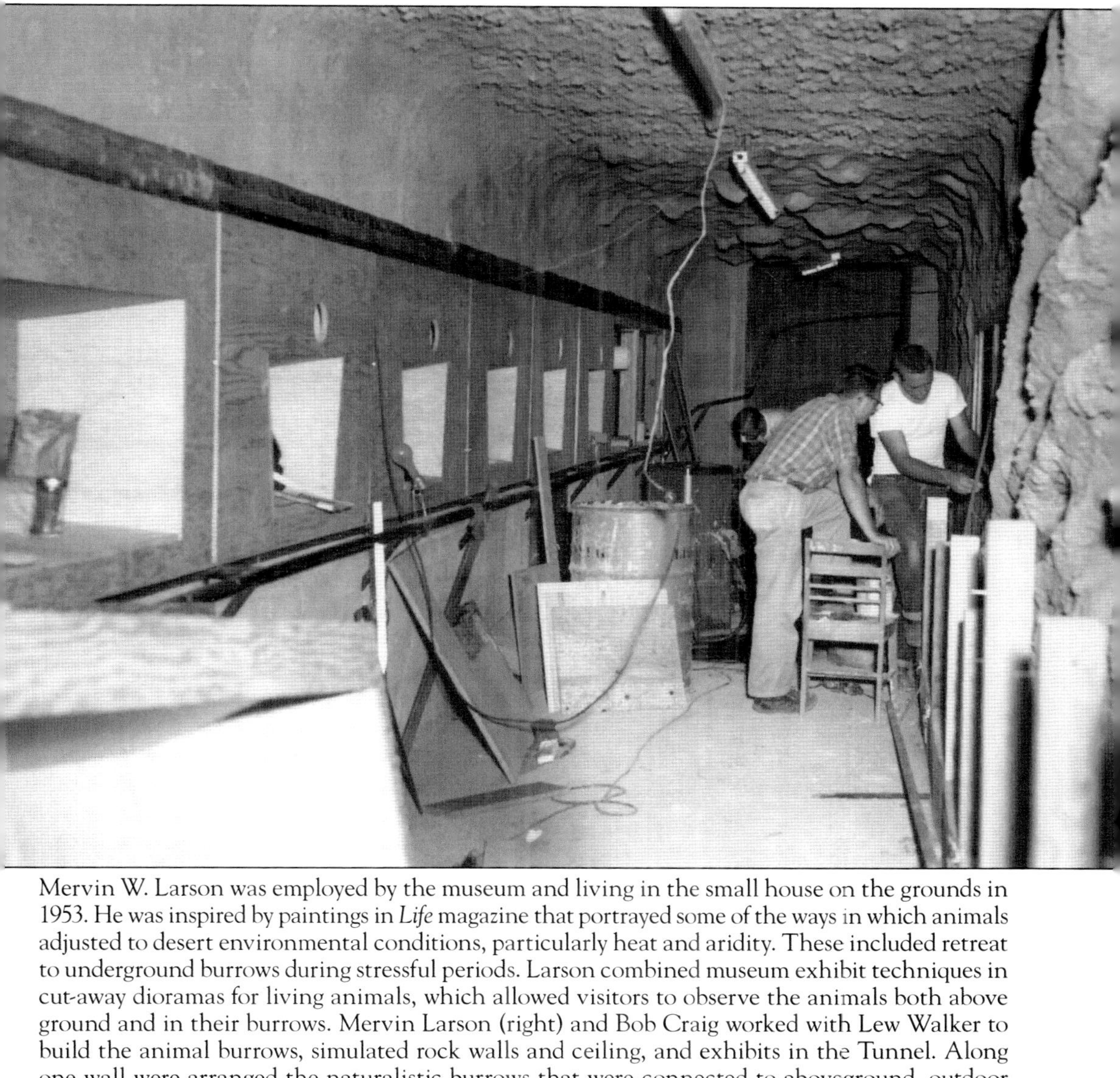

Mervin W. Larson was employed by the museum and living in the small house on the grounds in 1953. He was inspired by paintings in *Life* magazine that portrayed some of the ways in which animals adjusted to desert environmental conditions, particularly heat and aridity. These included retreat to underground burrows during stressful periods. Larson combined museum exhibit techniques in cut-away dioramas for living animals, which allowed visitors to observe the animals both above ground and in their burrows. Mervin Larson (right) and Bob Craig worked with Lew Walker to build the animal burrows, simulated rock walls and ceiling, and exhibits in the Tunnel. Along one wall were arranged the naturalistic burrows that were connected to aboveground, outdoor areas for the animals' use. On the other side of the Tunnel, plant roots, a cut-away ant nest, and other exhibits were located.

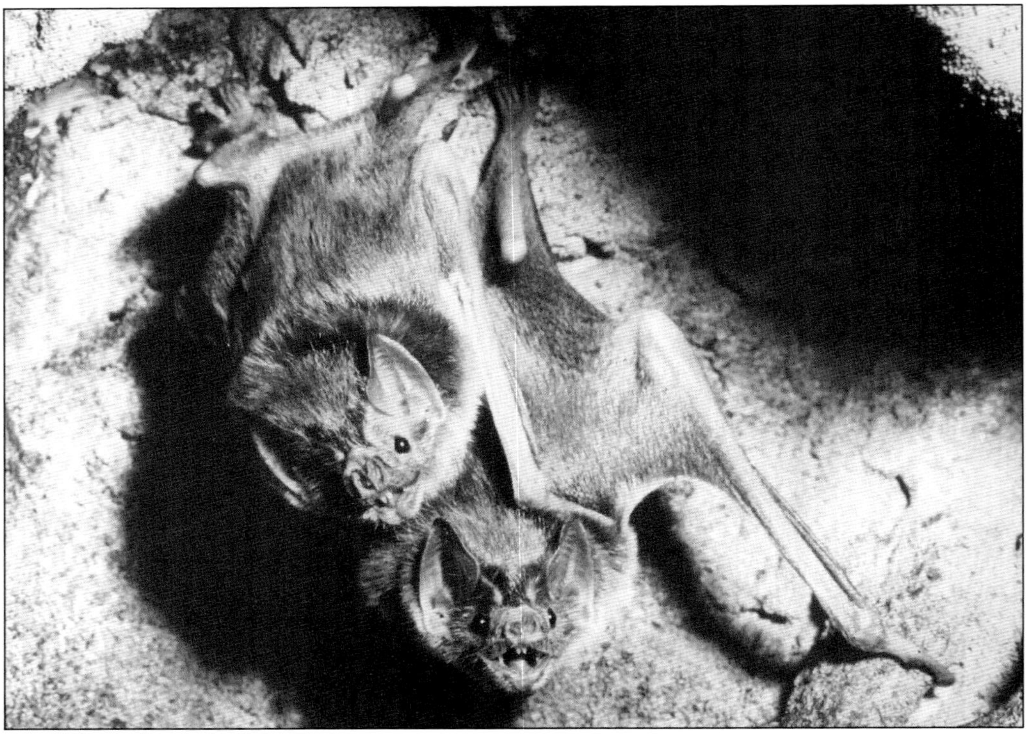

The bat cave in the Tunnel was a challenge, first to build with its elaborate formations and then to convince the bats to hang where they could be seen by the viewers. Fruit bats were housed in the cave. At one time, vampire bats (above) were housed in a nearby, separate section. The completed cave exhibit was outstanding, and doubtless a good many children viewing it grew up to be naturalists and avid cave explorers. (Above, photograph by Mervin W. Larson.)

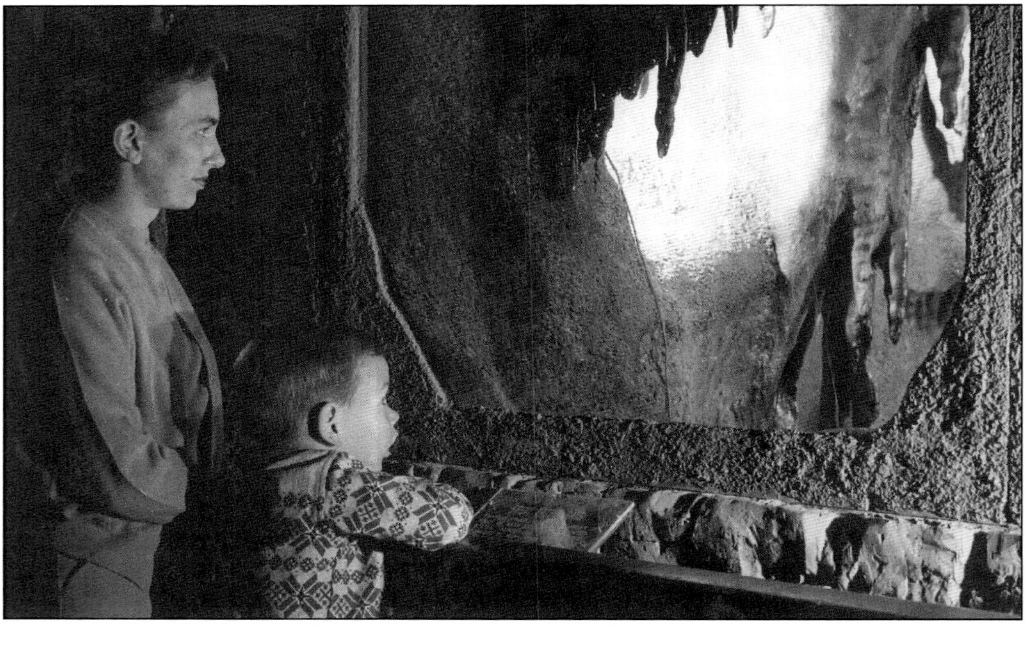

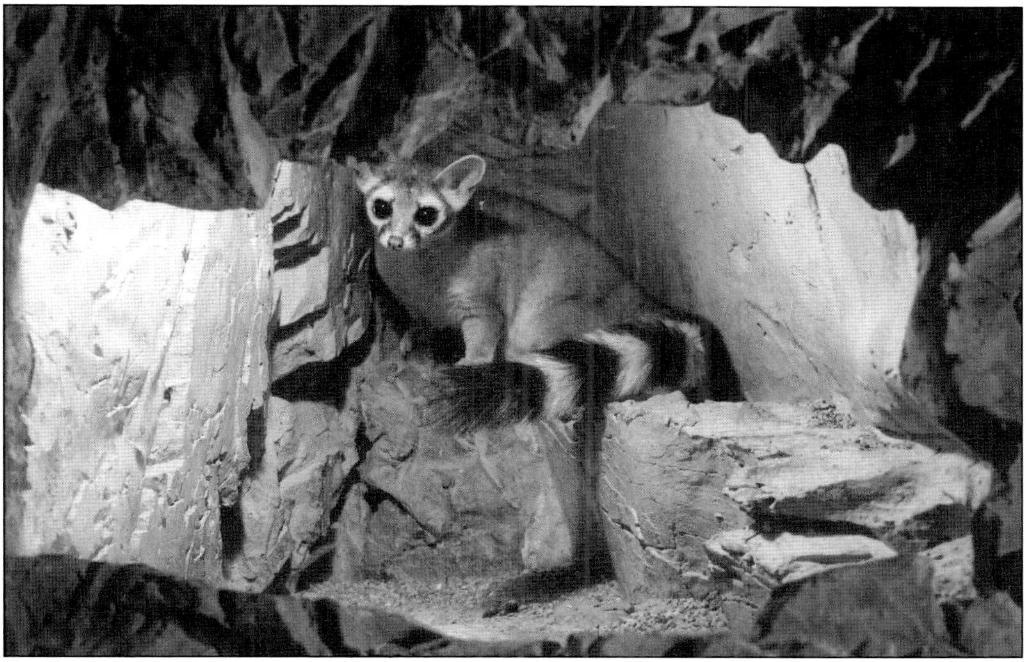

Ringtails are nocturnal, secretive, and not often seen in the wild, but nevertheless have been named Arizona's state mammal. This one posed in its burrow in the Tunnel, allowing visitors a close up look at its beautiful eyes and fluffy, black-and-white ringed tail. (Photograph by Wyatt Davis.)

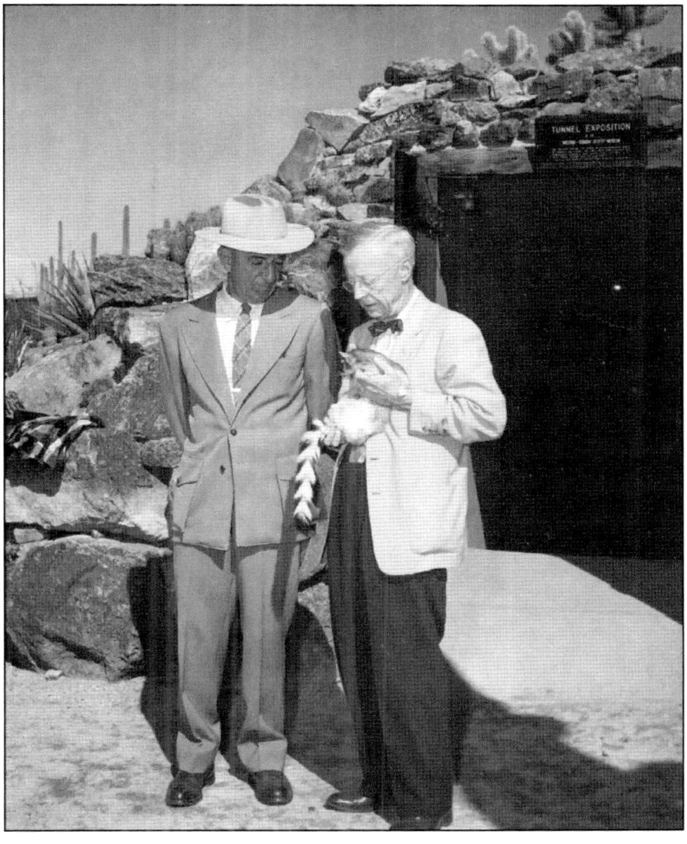

The Tunnel was kept in near darkness, but visitors could activate lights in each den. The animals soon became accustomed to the light and were not disturbed. One of the well known animals that was used in the Desert Ark program was the ringtail known as La Vaga, shown here being held by Arthur Pack (right) and observed by Carr. (Photograph by Hiram L. Parent.)

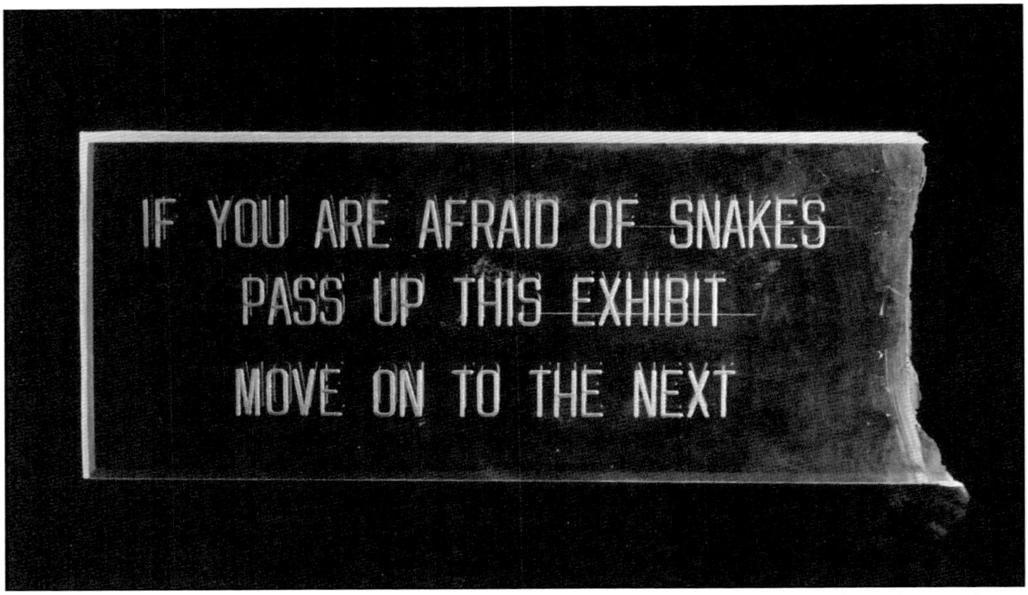

One very up-close exhibit in the Tunnel displayed one or more rattlesnakes only inches and one glass panel away from viewers. A sign placed just previous to the rattlesnake exhibit warned the faint of heart.

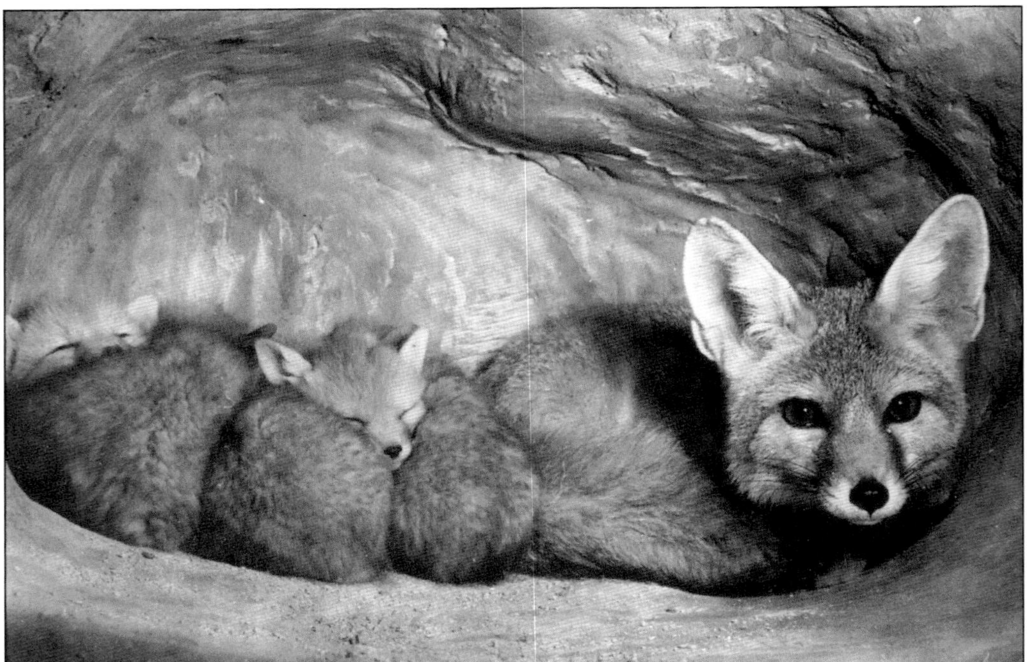
Several litters of kit foxes were born in the Tunnel over the years. This is the smallest of the foxes in the area. They can survive without free water, gaining moisture from the blood of their prey, and are therefore highly adapted for desert survival. Nocturnal, retiring to burrows during the day, kit foxes are seldom observed in the wild. What a privilege to be able to view this beautiful mother and her young.

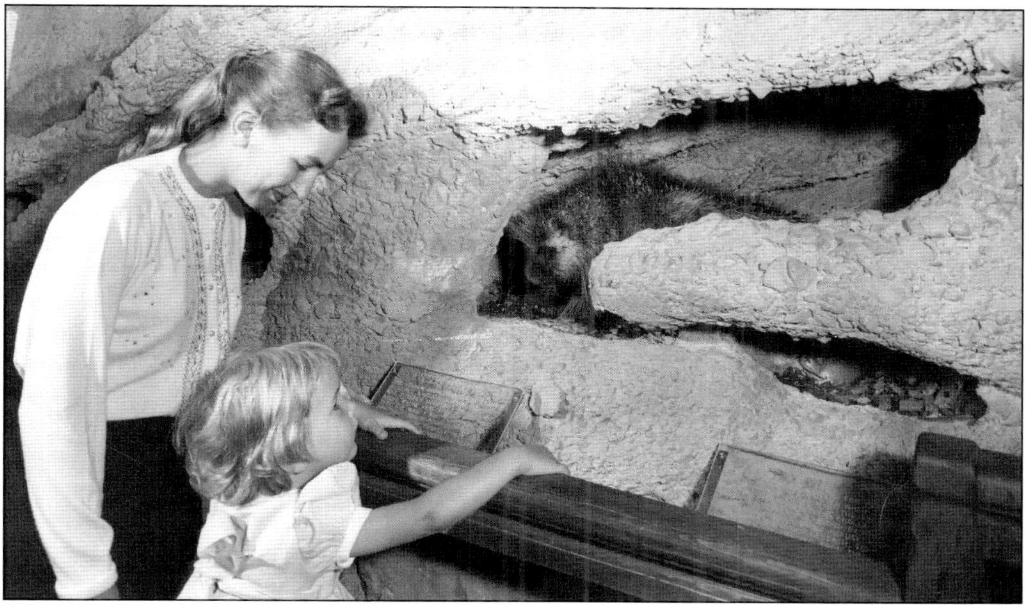

A porcupine and small girl go eye-to-eye in the Tunnel. These exhibits allowed a most unusual opportunity for human visitors to observe animals, separated from them by only inches. Some animal species were combined in the exhibits. In the cavity to the girl's right are two pack rats. A small passageway connects this burrow to that of the porcupine but prevents the larger mammal from entering the rats' private quarters.

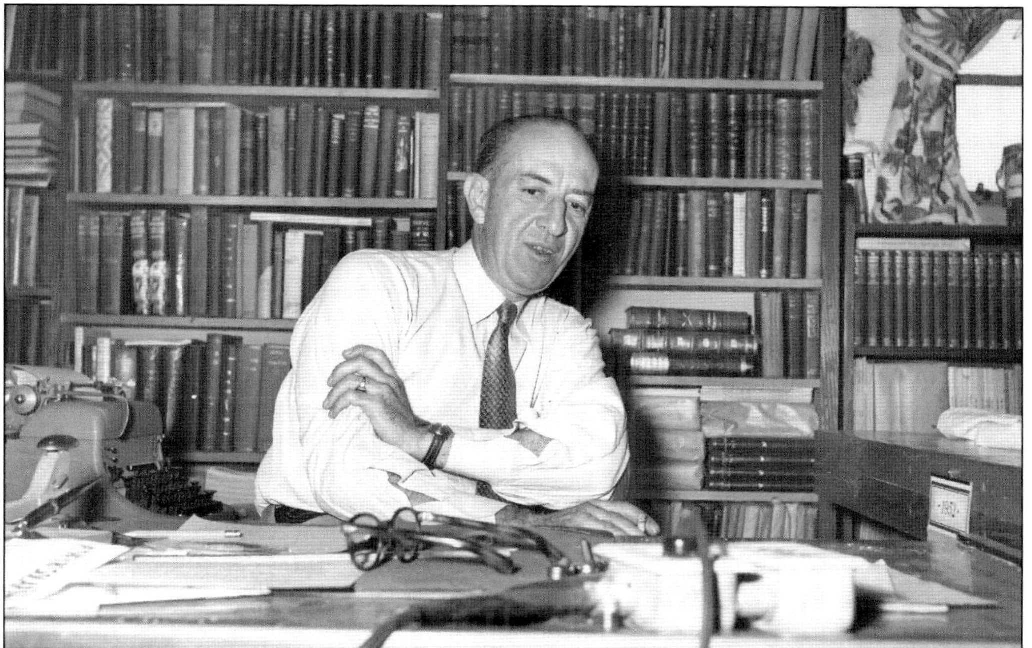

Bill Carr, progenitor of the Tunnel, sits in his office amid his library. An erudite and skilled naturalist, Bill Carr was an early pioneer in conservation through education in both his home state of New York and Arizona.

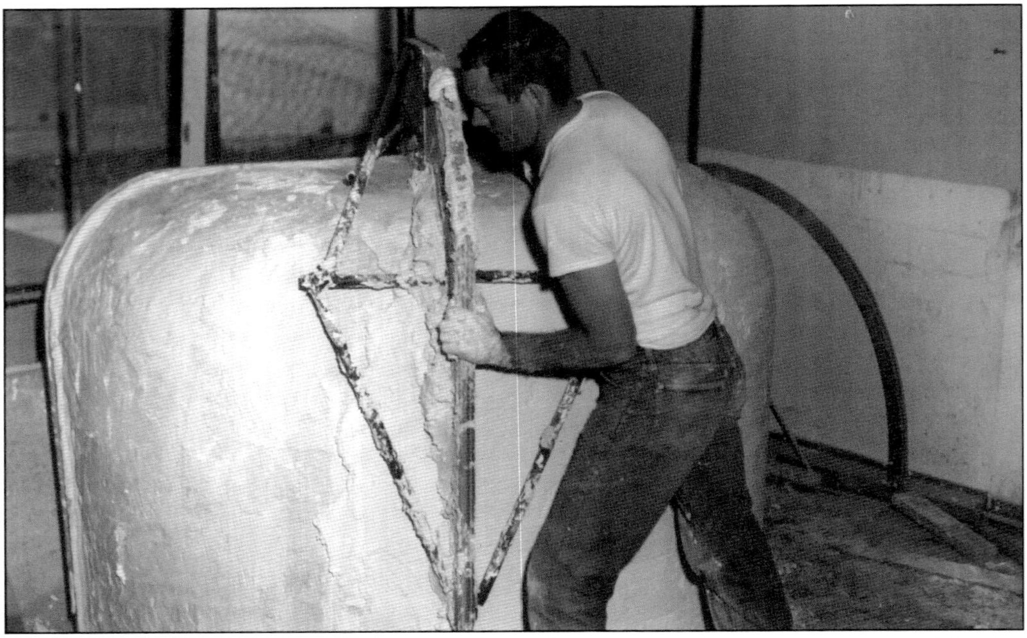

The exhibits in the amphibian room, opened in 1958, represented a major move to combine the traditional museum technique of painted backgrounds with techniques of live animal displays, an innovation developed by Merv Larson at the museum and soon copied by other institutions. Here Larson builds the mold for a "round back," which enclosed the back, top, and two sides of a diorama. A scene depicting a particular environment was painted inside the round back. A naturalistic setting was built on the flat bottom surface of the exhibit. The round back could be lifted hydraulically and accessed for easy maintenance of the exhibit. Five dioramas were the result. Like the Mexican river scene pictured, all have a painted background scene. This merges seamlessly into the foreground filled with constructed naturalistic rocks, water, and actual vegetation. Barely visible on the flat rock in the foreground is one of the diorama's living amphibian residents.

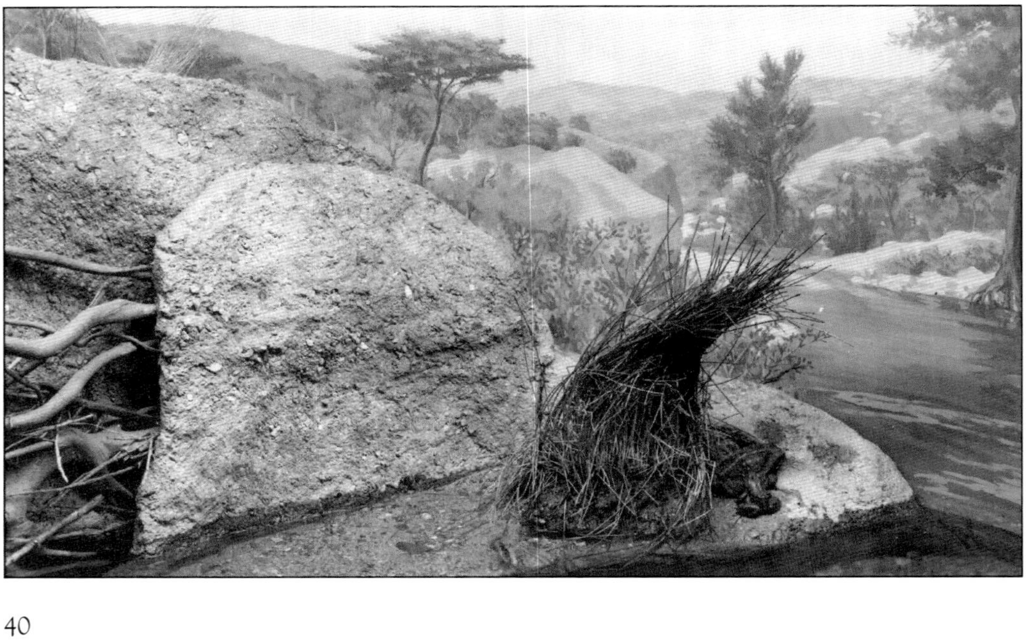

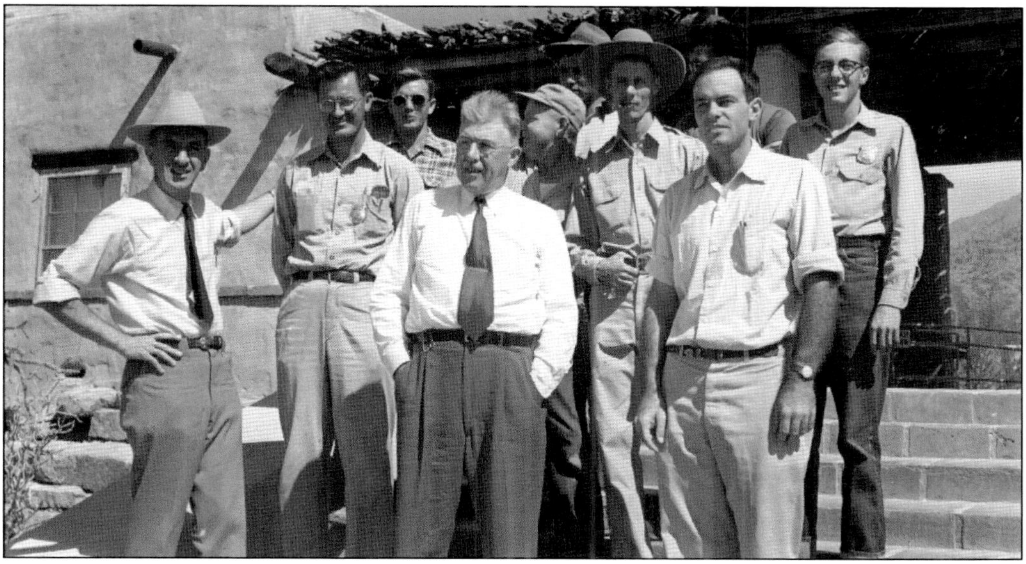

The early (probably 1954 or 1955) museum staff is pictured with a visitor (first row, center). Bill Carr (far left) and Bill Woodin flank him. Carr's hand is on the arm of Alan Blackburn (botanist), and behind him in sunglasses is Myron "Skip" Koth (large animal keeper). In the center with his face in profile is Rudy Johansson (maintenance and carpentry). Behind him in a hat is an unknown employee. Facing the camera, wearing a big hat and smoking a pipe, is Floyd Miller (animal keeper). Above Woodin's left shoulder is Merritt Keasey (small animal keeper).

This young jaguar was born in captivity and acquired by the museum. While young, it was raised in Lew and Melanie Walker's household when they lived at the museum. The cat weighed 40 pounds when it arrived and began to grow rapidly. The day came when Melanie suggested strongly that it was time for the young one to leave home and join the other museum animals—understandable when looking at its claws.

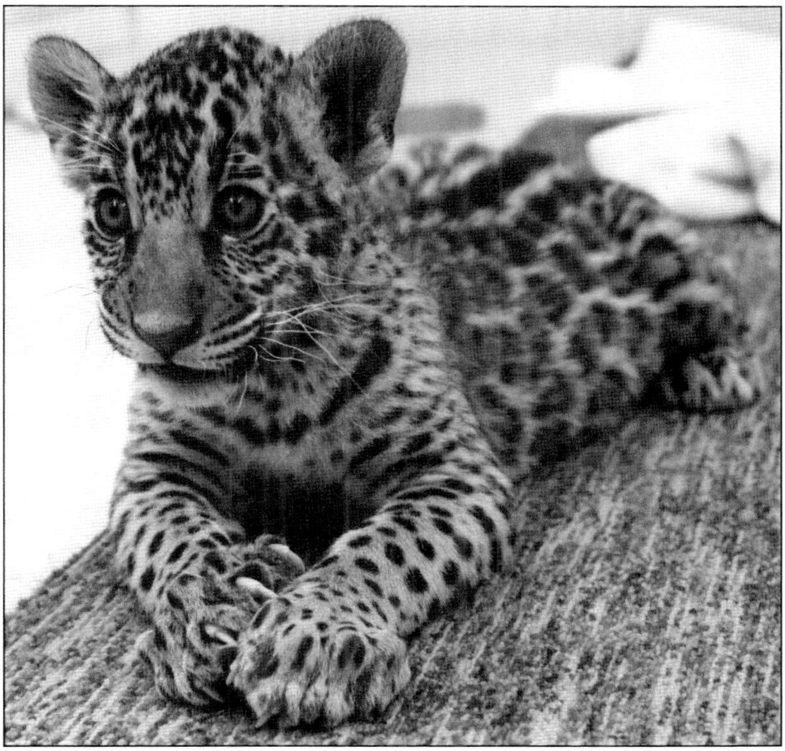

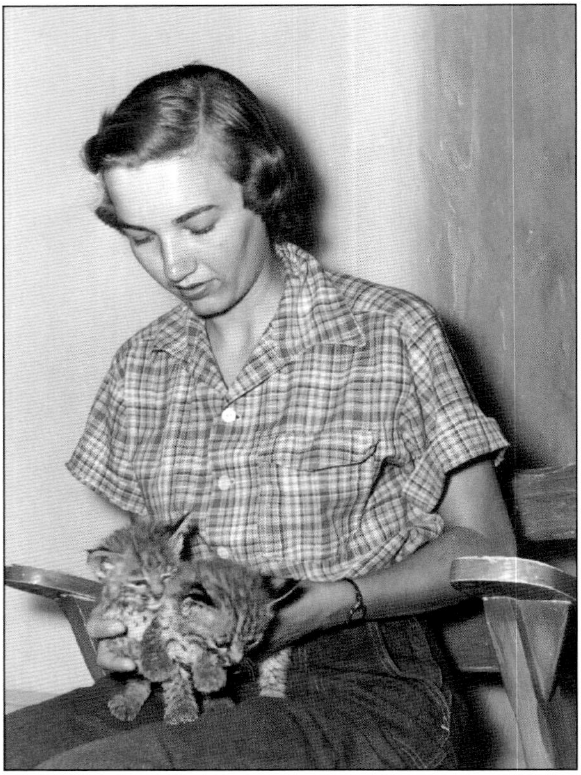

In the earliest days of the museum, there were no wildlife rehabilitation facilities in Tucson. Orphaned and injured animals were frequently brought to the museum, and the employee and his family who lived in the small house up by the stable at the time often became their caretakers. Pictured with Peggy Larson, these two orphaned bobcats, Whiskers and Buttons, grew up to be fine, friendly cats but could never be reintroduced to the wild, having been raised in captivity.

Mountain lions have always been highly regarded at the Desert Museum and serve as the museum's icon. This is Suzy, the museum's first lion. She was a beautiful animal but reserved. There were hopes for baby lions, and a male was obtained, but Suzy refused to accept him as a friend or mate and the pair produced no offspring. (Photograph by Marvin H. Frost.)

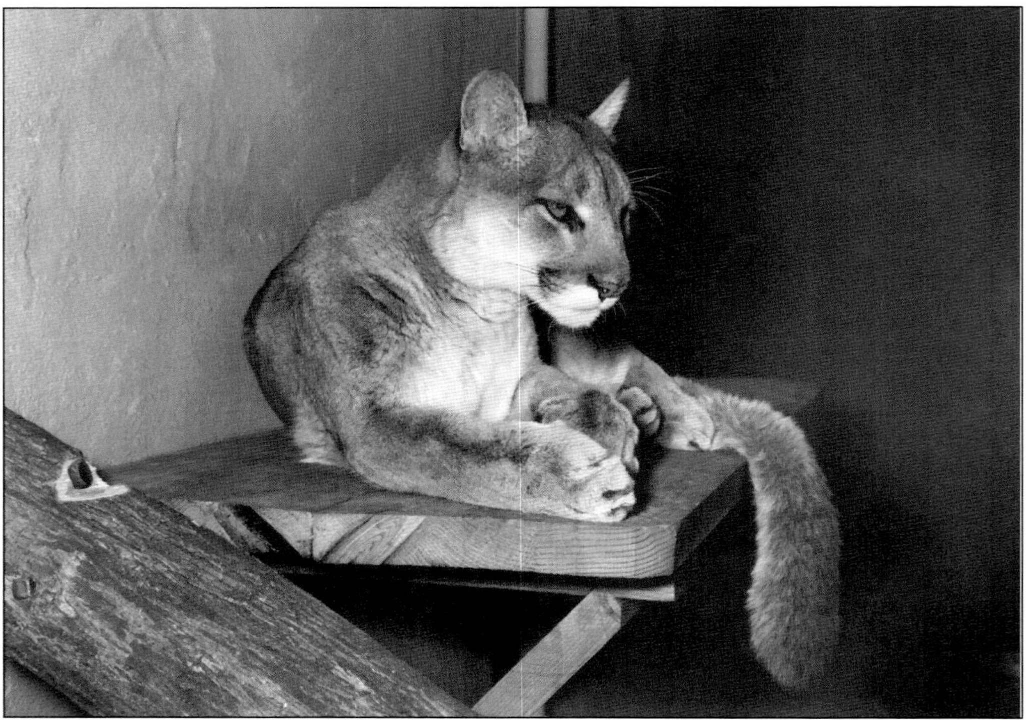

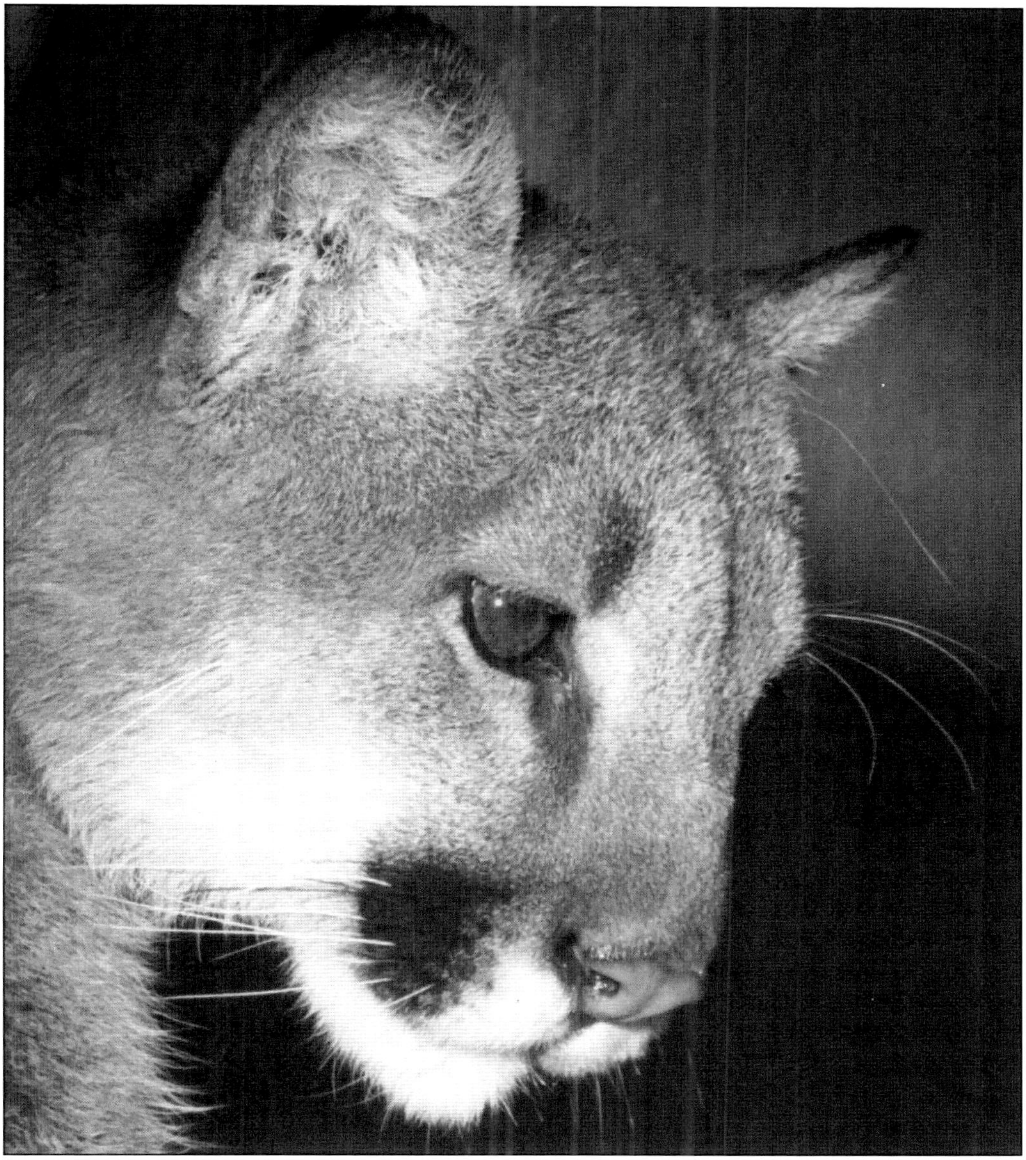

The male lion that Suzy rejected had been raised among people in California before coming to the Desert Museum. He was young with a playful personality, quickly charming the staff, visitors, and community. Arriving on George Washington's birthday in 1953, he was named George. Bill Carr added L. for Leo (the lion). George loved to have his keepers scratch his ears through the openings in the chain link front of his enclosure. When happy, which was any time he was receiving attention, he purred loudly, yapped at those he knew, and quickly destroyed any object meant for his entertainment left in the enclosure. Staff members had been writing a weekly column about museum activities that was published in the *Arizona Daily Star*. George took over the authorship (with a little help from Carr, Gras, Larson, and others). With his own byline, George began to receive letters from his fans—at one time up to 100 letters a month. He became so well known in town that when a child's letter to George came addressed simply as "George, Tucson, Arizona" it was correctly delivered.

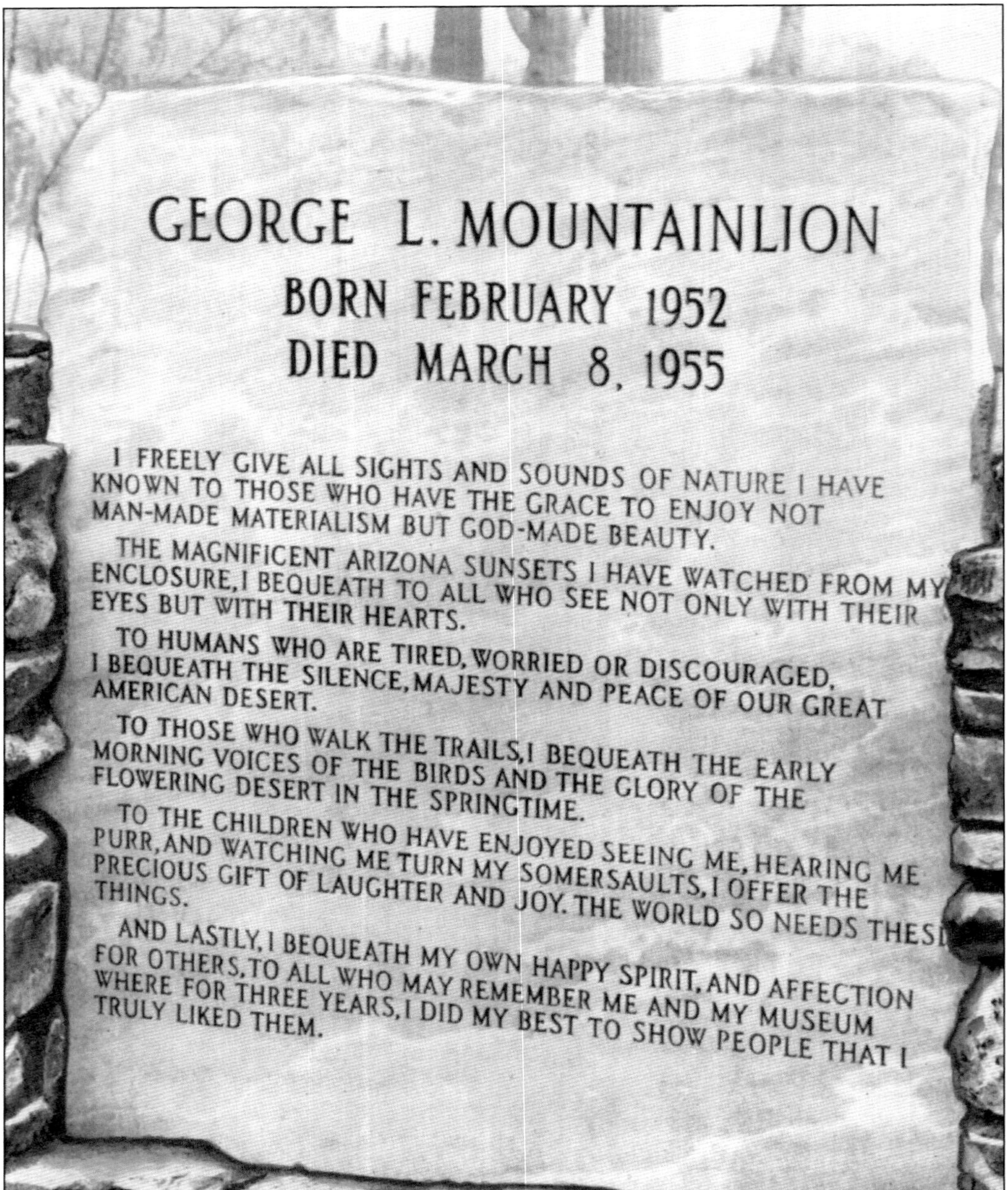

Two years later, George became ill with liver disease. As his condition became critical, the community worried. The local newspapers carried daily bulletins regarding his condition. An appeal was made to local hunters for the donation of venison from their freezers for George's faltering appetite. There were no specialized mountain lion veterinarians in Tucson, so Lew Walker drove the big cat to the San Diego Zoo seeking help. Despite the zoo's dedicated efforts, George died there. The Tucson newspapers carried front-page headlines with the news. George was buried early one morning at the museum just across the walkway from the front of his enclosure. Carr wrote George's will, which was engraved on the stone that marked his grave. Today the stone forms a part of the museum's history wall.

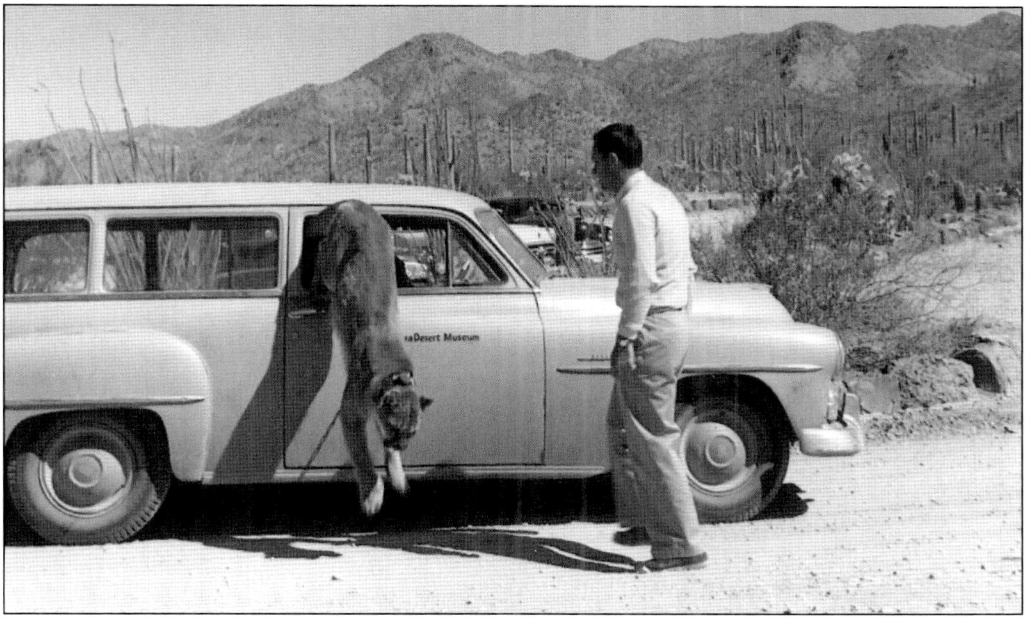

George L. Mountainlion II arrived at the museum soon after the death of George I. He had been raised by the Willis Parker Animal Agency in California, and Lew Walker traveled to Berkeley to pick him up. Here Bill Woodin greets the new arrival. George II proved to be as likeable as George I, and he soon delighted visitors with both his personality and his authorship of the weekly newspaper column.

In the early years of the museum, the families of more than one staff member had their Christmas card pictures taken with a special museum resident. Here Merv, Peggy, Lane (left), and Lynn Larson pose with George II. The parents smile brightly while the father grips George II, the son and daughter radiate apprehension, and the mother smiles cluelessly as George II tries to take a swift swipe at her son.

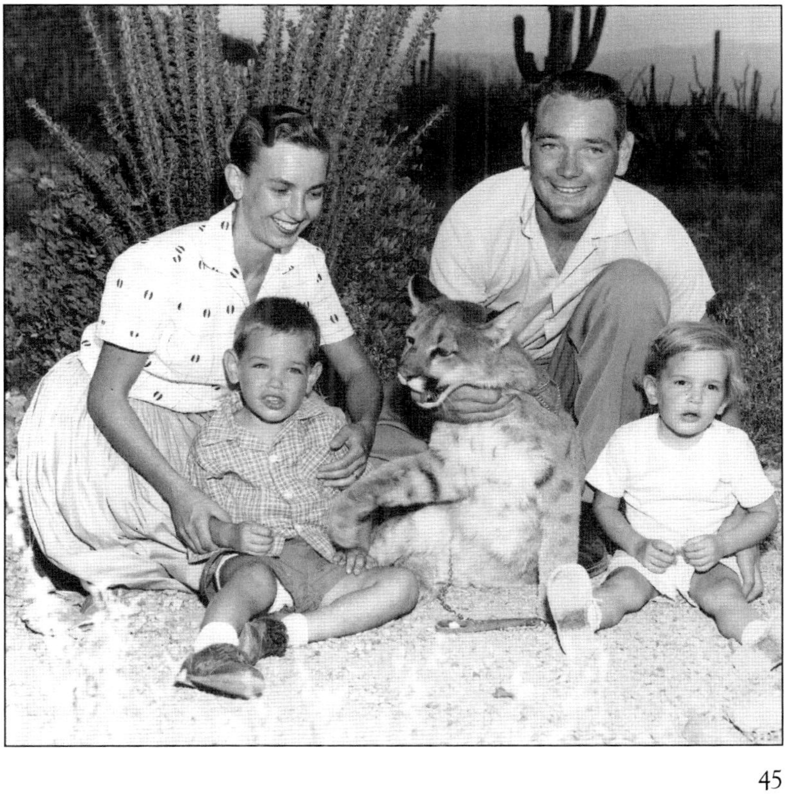

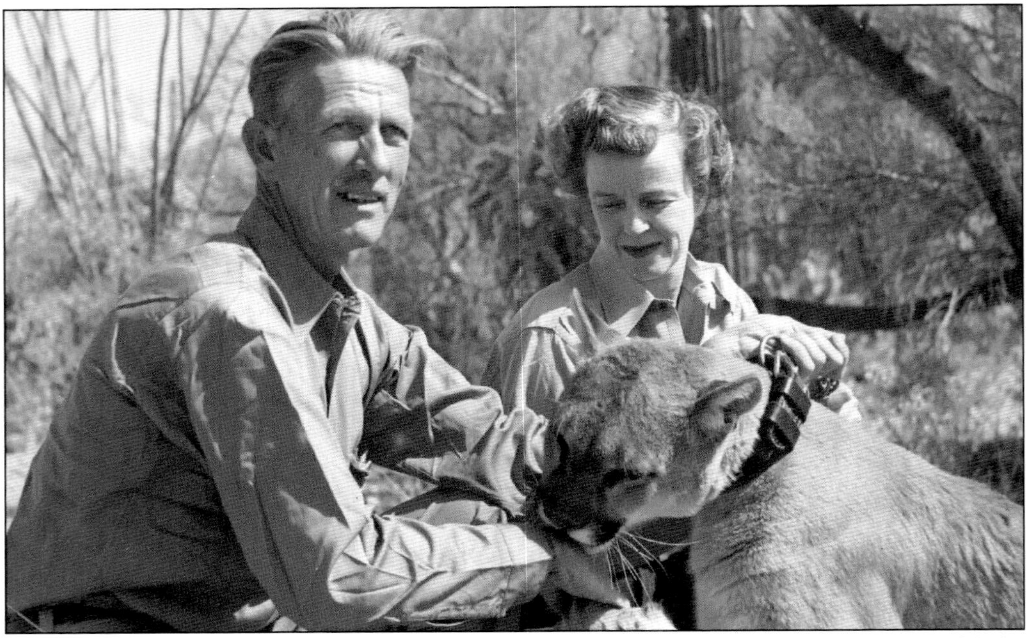

George L. Mountainlion II could be taken out and walked on the grounds, much to the delight of both George II and the museum's visitors. He became the subject of innumerable photographs. Here Melanie Walker observes the beautiful adult cat. George II appears to be holding hands with Lew while also licking that hand with his super-rough tongue.

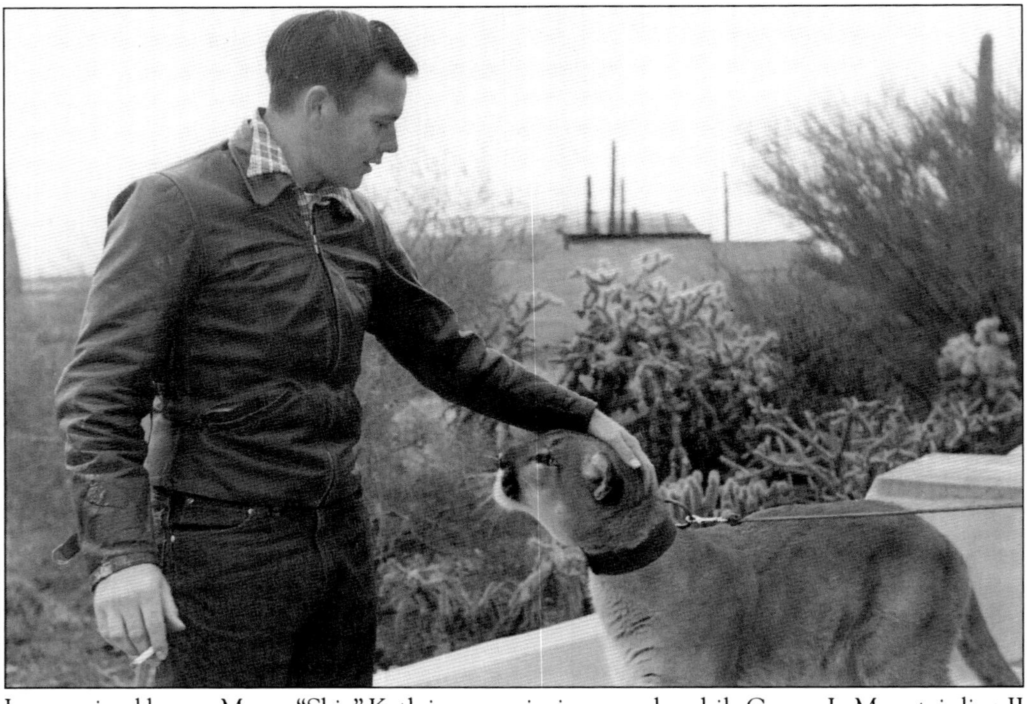

Large animal keeper Myron "Skip" Koth is seen enjoying a smoke while George L. Mountainlion II is benefitting from a pat on the head as they make their rounds of the museum grounds.

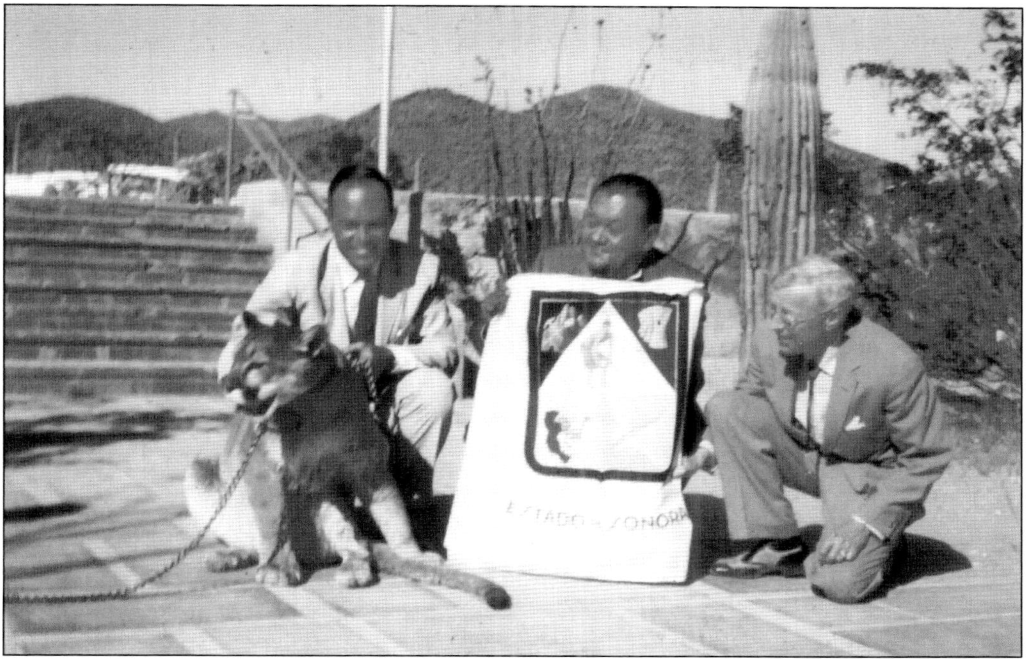

Mexican consul Cosme Hinojosa is shown holding the unofficial coat of arms of Sonora at the Arizona-Sonora Desert Museum with the backdrop of the Tucson Mountains. The coat of arms bears the words *Estado de Sonora* (State of Sonora). George L. Mountainlion II is seated in front of director Bill Woodin. Arthur Pack, president of the board of trustees, is seen on the right.

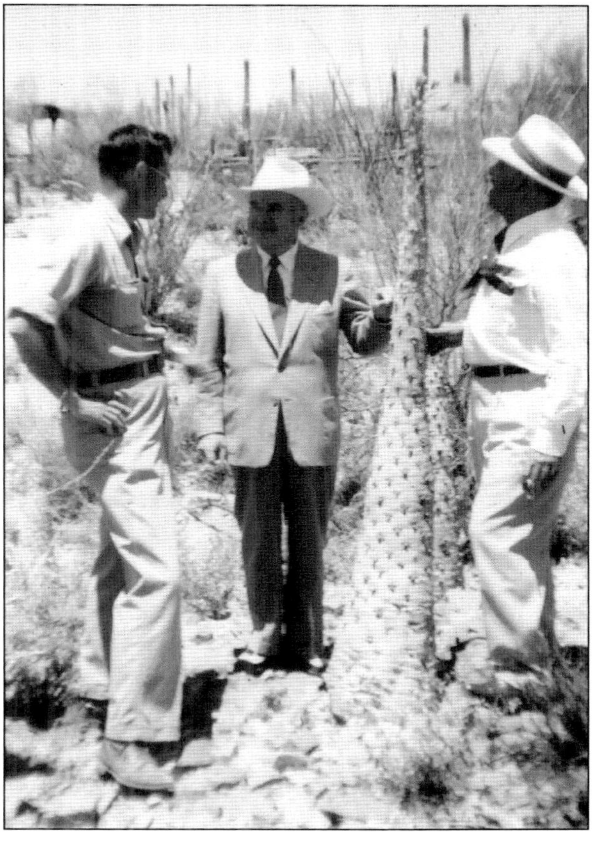

Oscar Soto Maynez (center), governor of the state of Chihuahua, Mexico, was photographed on the museum grounds during the early 1950s. The governor and his aide, along with Alan Blackburn (left), the museum's botanist, discuss the properties of the boojum tree. The museum has always worked closely with governmental officials in Mexico, and the governor of the state of Sonora has often been an ex-officio member of the museum's board of trustees.

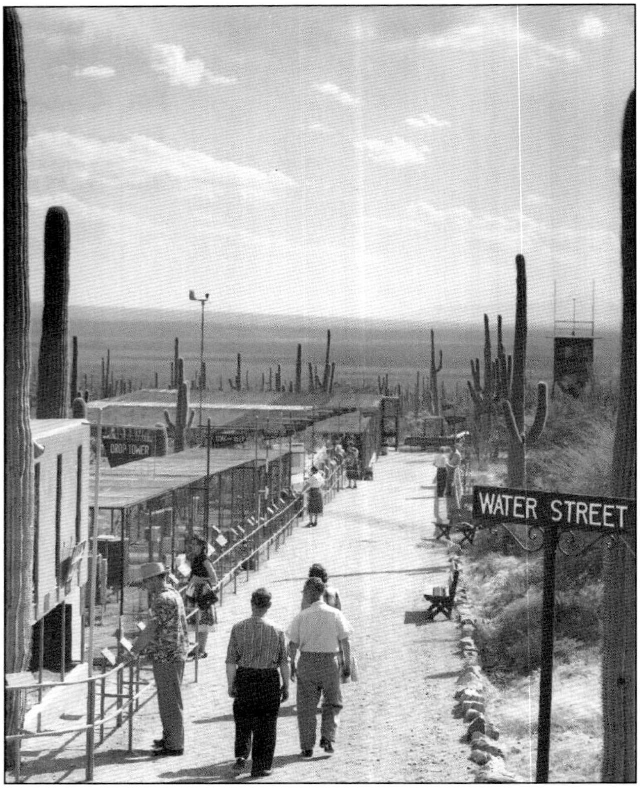

The museum was acting far in advance of today's heightened concern for the availability of water when Water Street was dedicated in 1958. Carr wrote, "We have attempted to demonstrate how mankind learns about the influence of water upon his land and life, and to tell why it is absolutely essential." Carr, with the assistance of many hydrologists, institutions, scientists, and museum personnel, planned and built the exhibits. The Pack Foundation provided funds to build Water Street and presented this major exhibit as a gift to the museum. Water Street included displays on rainfall, evaporation, infiltration, moisture content, different types of plant cover, the effect of rain, results of overgrazing, and many other related subjects.

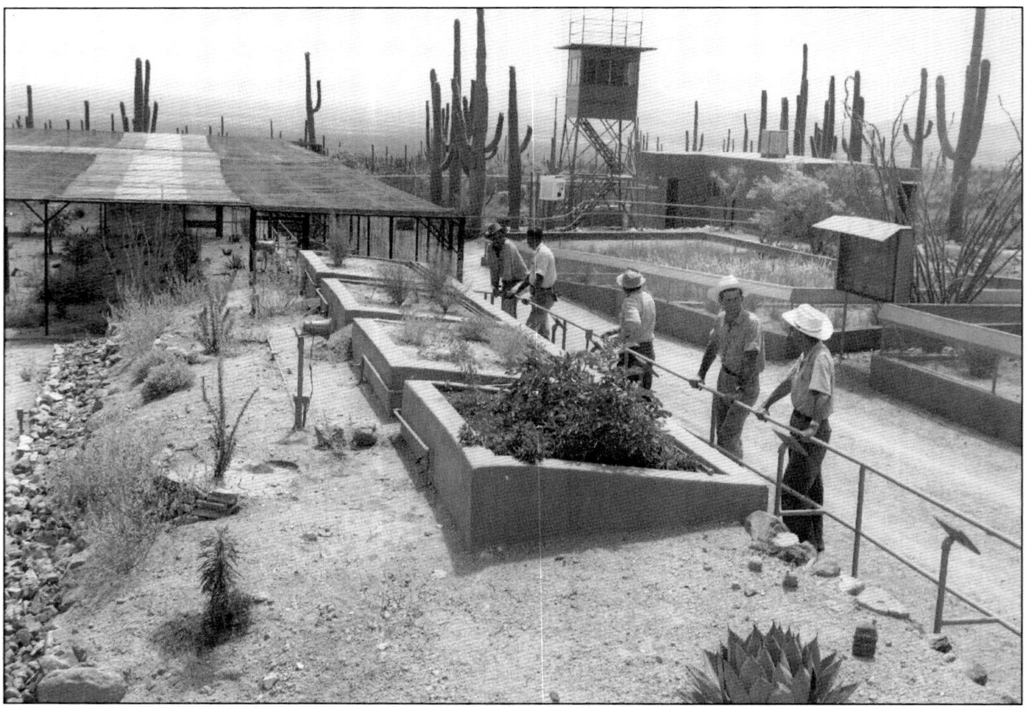

Exhibits on Water Street often included scientific or electronic equipment. Some could be visitor activated. The equipment, labels, and living plants and animals told the story of water and its relationship to all living things. Here a staff member services evaporative pans, one of which demonstrates that use of a harmless film on the water's surface slows down evaporation 50 percent compared to that of the untreated water in the adjoining pan.

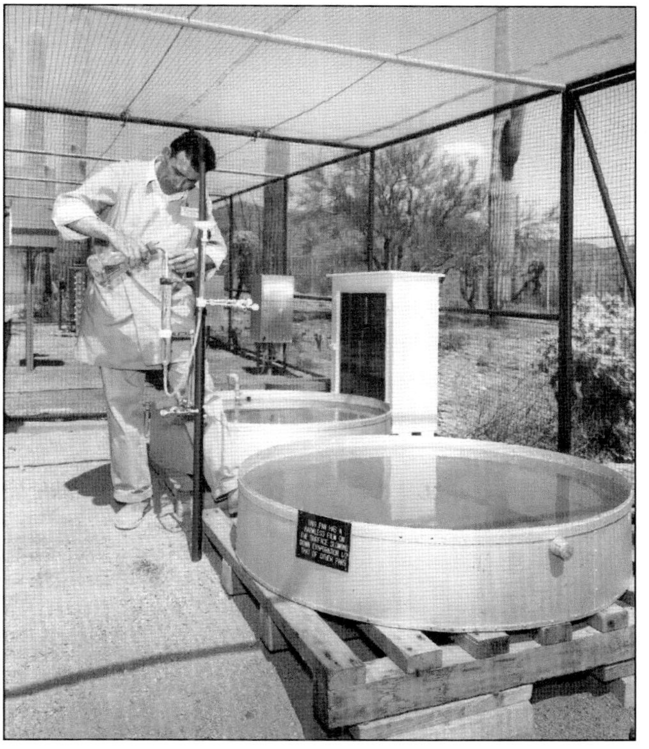

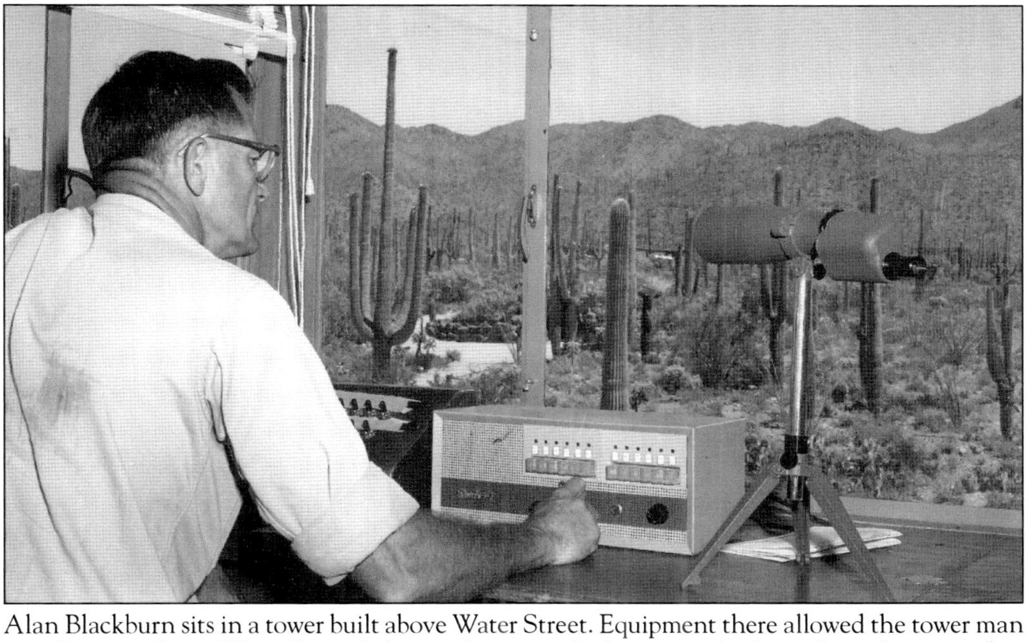

Alan Blackburn sits in a tower built above Water Street. Equipment there allowed the tower man to talk to people on Water Street below, and conversely there were facilities along the street that allowed visitors to ask him questions or to relay their comments. With a telescope, the tower man could closely observe individuals who might have questions or engage in conservation with those who evidenced an interest in a particular exhibit. (Photograph by Wyatt Davis.)

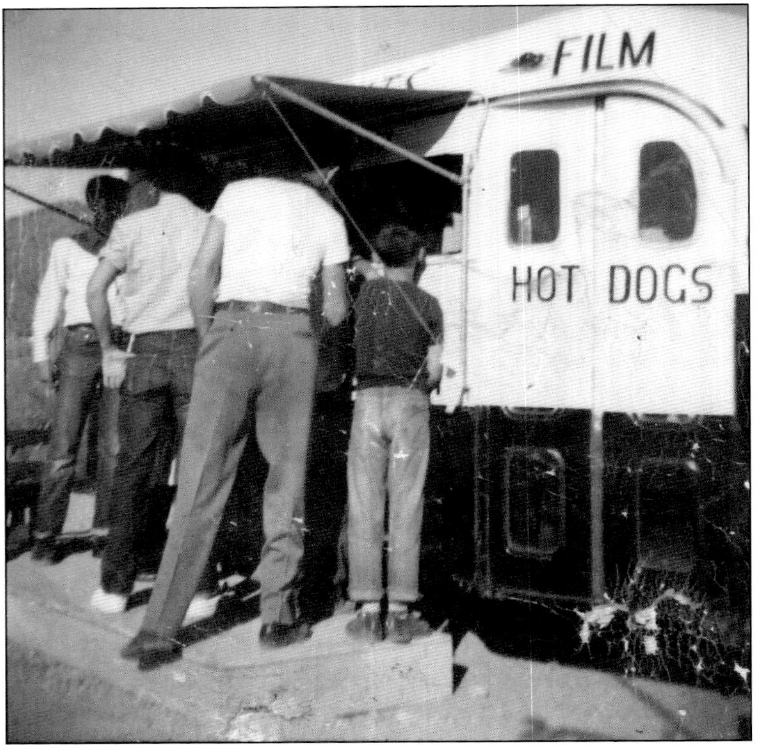

In 1952, there was no food or drink concession at the museum. One day, sometime after the opening, a converted school bus appeared in the parking lot and the Pat Hogan family began selling cold sodas, hot dogs, hamburgers, and camera film. Carr quickly made a permanent arrangement with the bus owners. Soon thereafter, the Hogans' food concession moved into more permanent quarters in the east wing of the museum building. (Courtesy of Hogan family.)

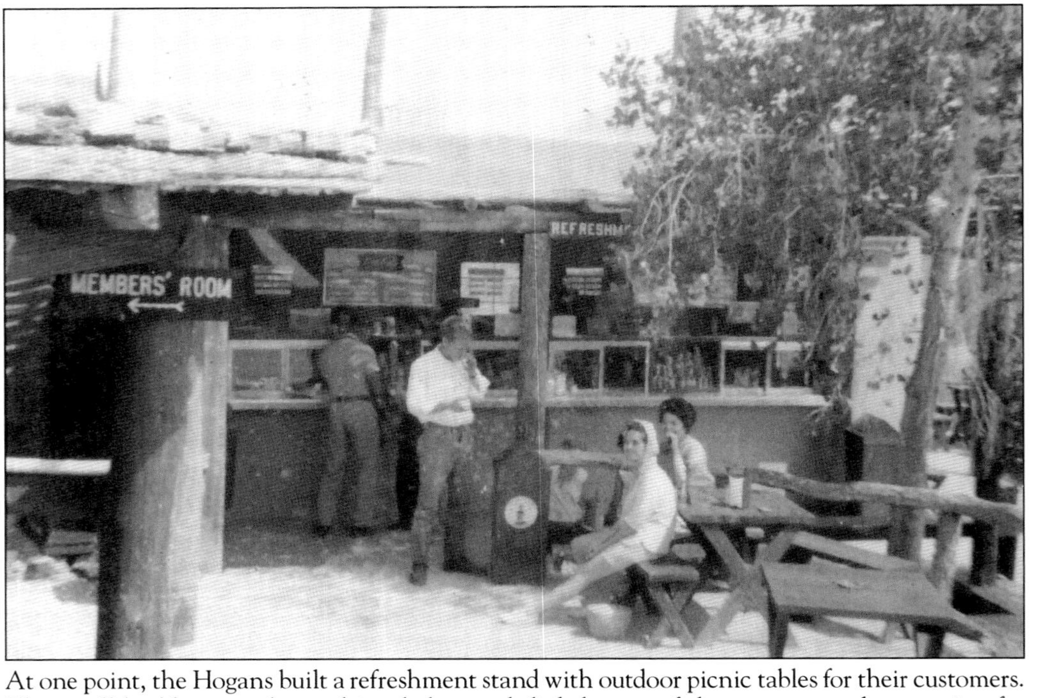

At one point, the Hogans built a refreshment stand with outdoor picnic tables for their customers. The small building was located just below and slightly east of the current outdoor seating for Phoebe's coffee shop.

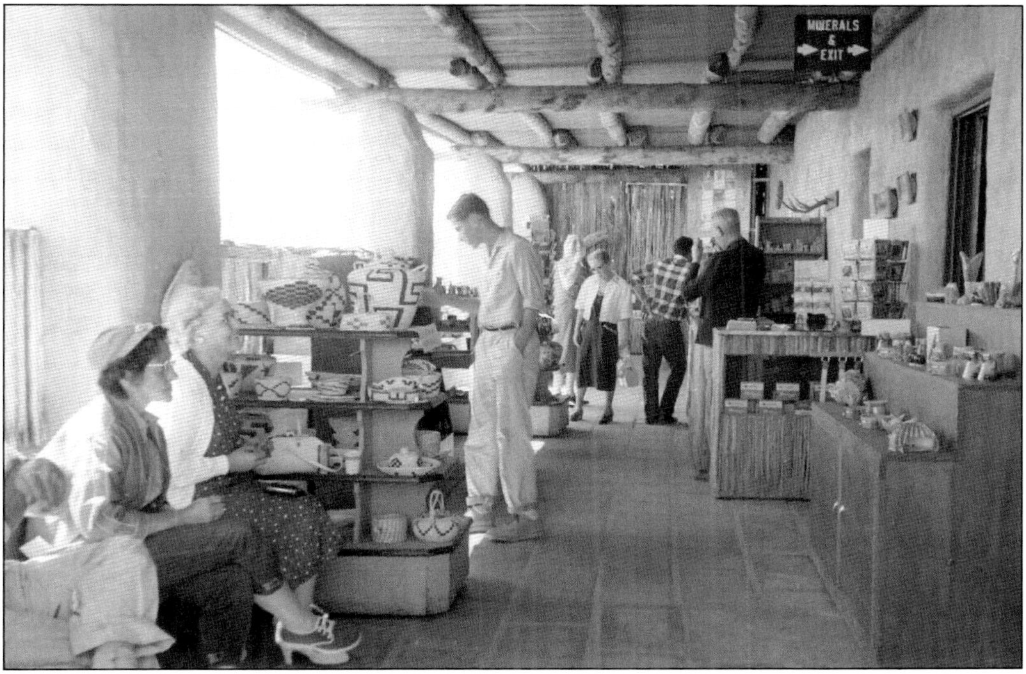

The first gift shop was located on the partially open porch on the south side of the east wing. Merchandise was varied but strong on postcards, small living cactus plants, and poor, dead scorpions and tarantulas permanently entombed in clear plastic paperweights. (The museum soon banned the dead wildlife paperweights.) There was a nice display of quality Papago (Tohono O'odham) baskets. A sign points right to a small mineral display housed inside.

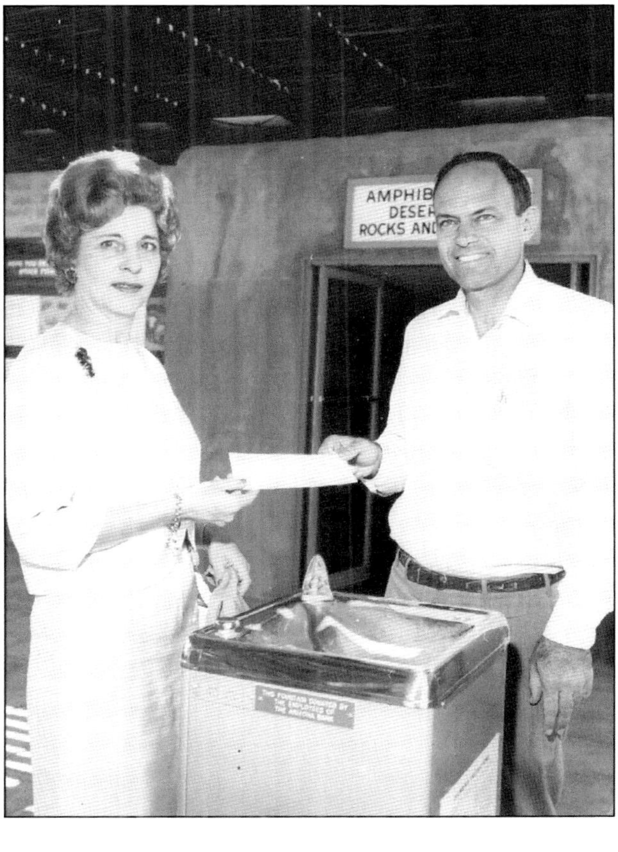

Tucson community businesses and individuals were generous with their gifts. Here an employee of the Arizona Bank presents a check to director Bill Woodin (right), making possible the purchase of this drinking fountain, which is already labeled as the bank's gift. Drinking fountains around the grounds have always been much appreciated by museum visitors on warm desert days. (Photograph by Mervin W. Larson.)

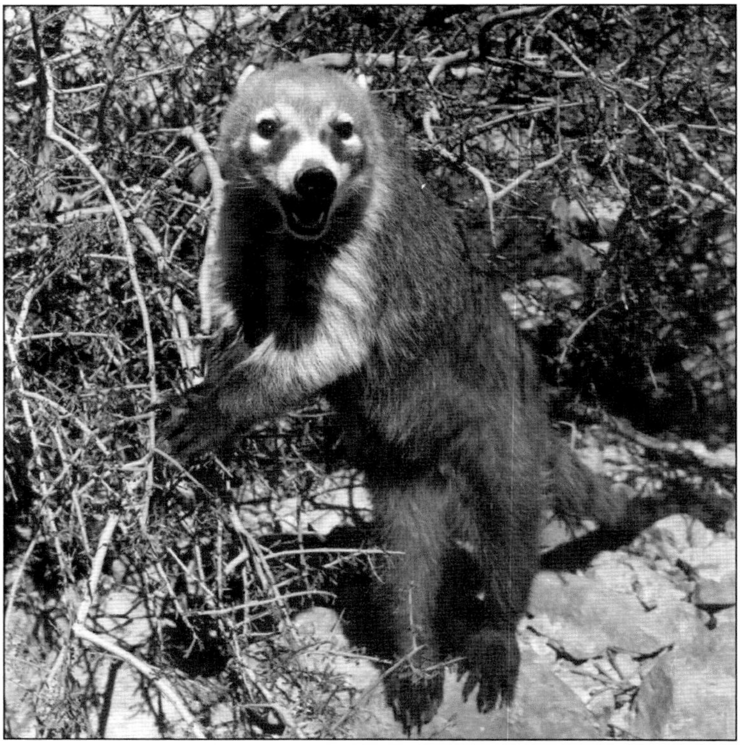

Originally from Mexico, the coatimundi is related to the raccoon. Coatimundis are found in Mexico and along the borders of Arizona, New Mexico, and Texas. They are social animals, and vocalize with snorts, grunts, and chatters. Coatis spend their days foraging for food, which includes lizards and a variety of fruits, including prickly pears. They are excellent climbers and sometimes spend the night sleeping in trees.

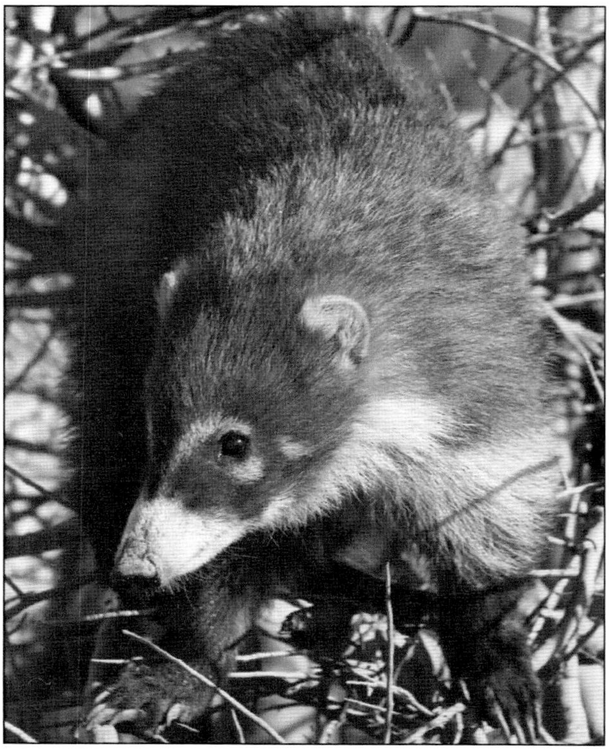

Contributions in a donation box helped make possible the early open, moated coatimundi exhibit. These diurnal animals live in social bands of up to 30 members. The facemasks of coatis are similar to that of raccoons and their tails reminiscent of ringtails. Coatis, however, are a distinctive species, and when traveling, they carry their two-foot-long tail held vertically.

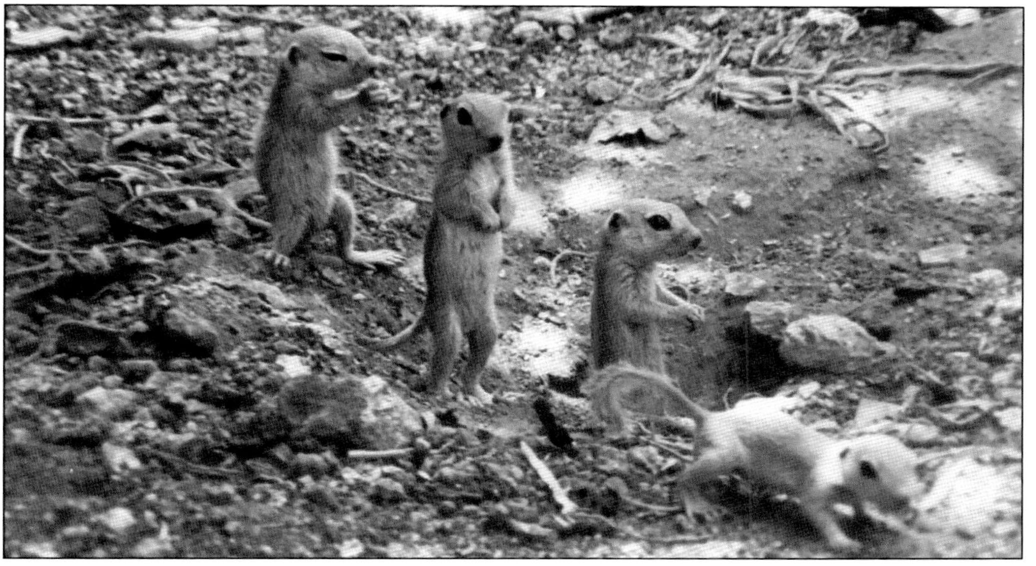

Not all animals at the Desert Museum are confined. The grounds attract many birds, insects, lizards, rabbits, and other small mammals. Here a group of round-tailed ground squirrel siblings have emerged from underground. These squirrels live in small colonies. They hibernate during the winter, emerge in spring to eat new green growth, estivate through the heat and aridity of summer, then reemerge when summer rains produce new plant growth.

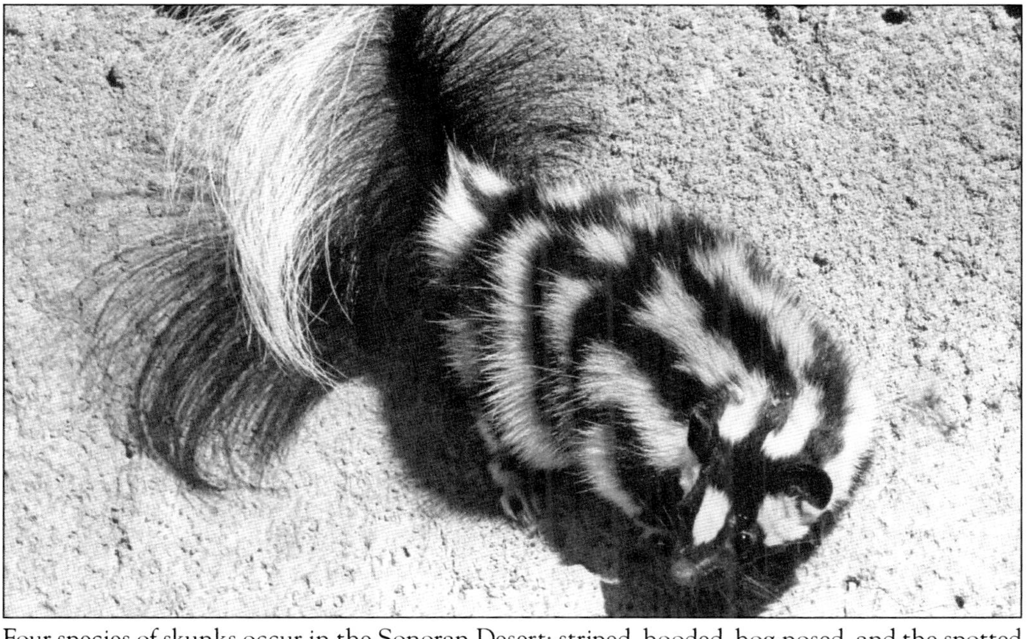

Four species of skunks occur in the Sonoran Desert: striped, hooded, hog-nosed, and the spotted skunk, pictured here. The spotted is the smallest of these; its white coloration resembles spots, rather than stripes. It is the only one that can climb trees and, unlike the others, does a handstand before spraying. Before the museum opened, the building was partially open during renovations and skunks moved into several lower kitchen cupboards. Reportedly they and the workmen lived in mutual respect and harmony.

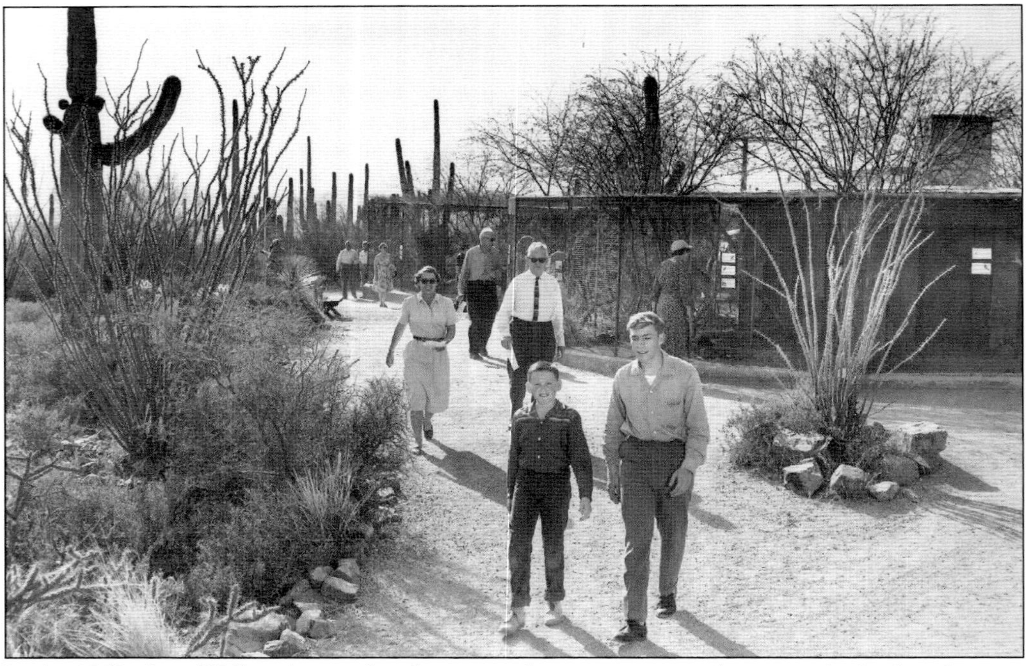

In the 1950s, Lew Walker designed and oversaw the construction of two circular bird enclosures. These could be serviced from the center area; the birds were readily visible and they lived comfortably in summer heat with their own evaporative cooler. These "circulars" remained in use for over half a century.

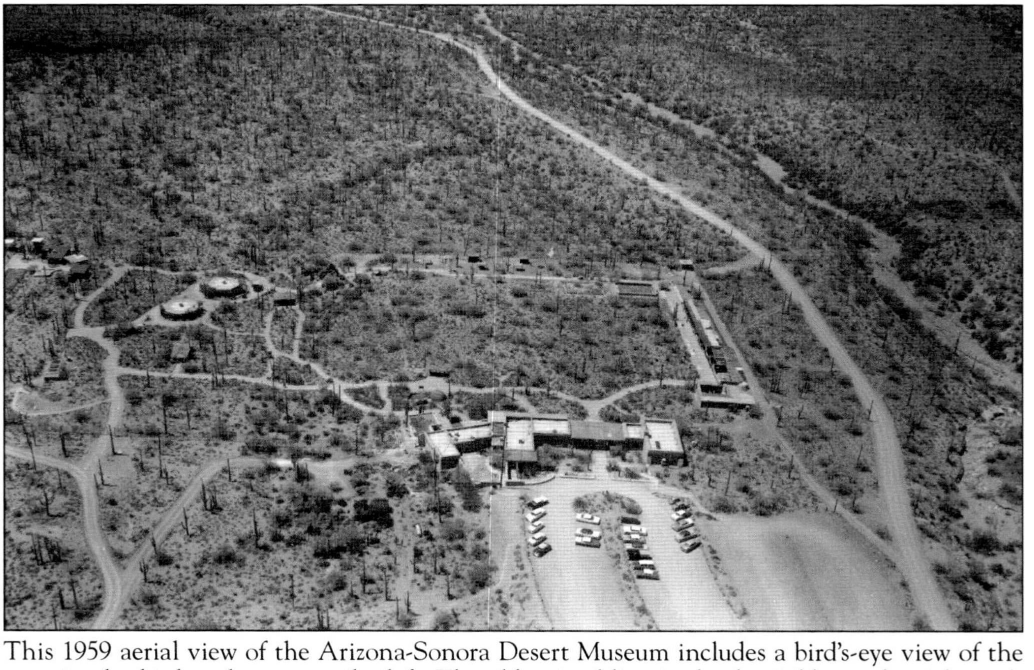

This 1959 aerial view of the Arizona-Sonora Desert Museum includes a bird's-eye view of the two circular bird enclosures on the left. The old animal line is clearly visible on the right with the original Tucson Mountain House building in the foreground.

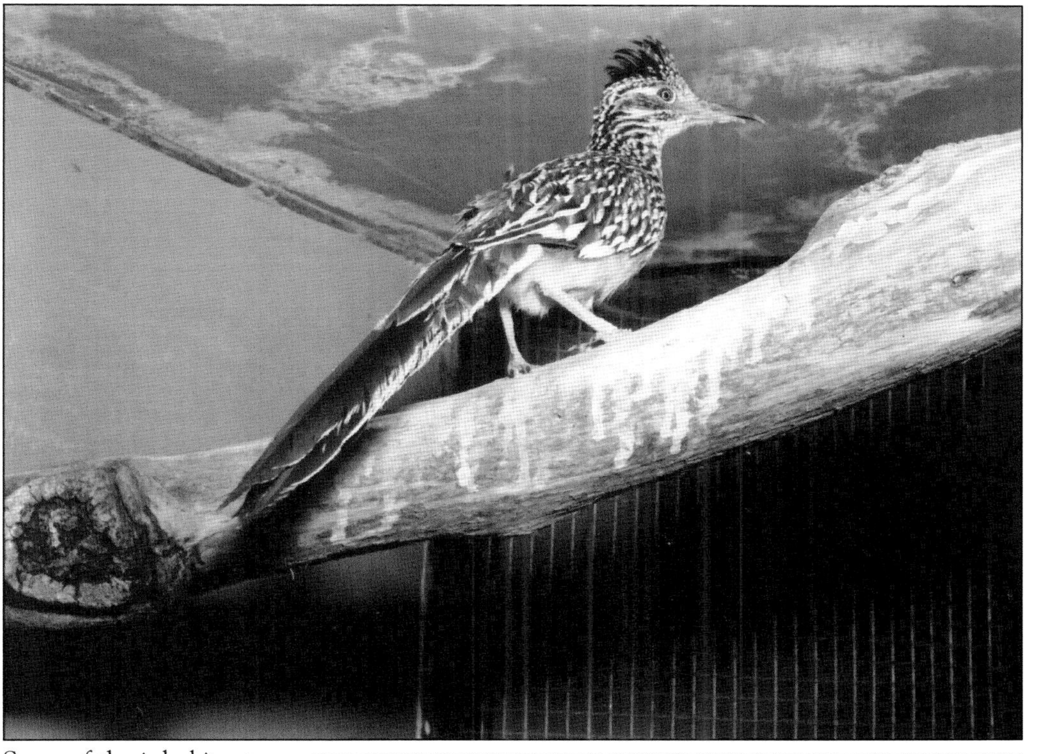

Some of the inhabitants found in the circular bird enclosures included the roadrunner, seen above. Renowned for feeding on rattlesnakes, the roadrunner is as fast as it is agile, attaining speeds up to 17 miles per hour. Additional birds, seen at right, include the American bald eagle and the great horned owl. Both of these raptors are active in the skies above southern Arizona. (Both photographs by Mervin W. Larson.)

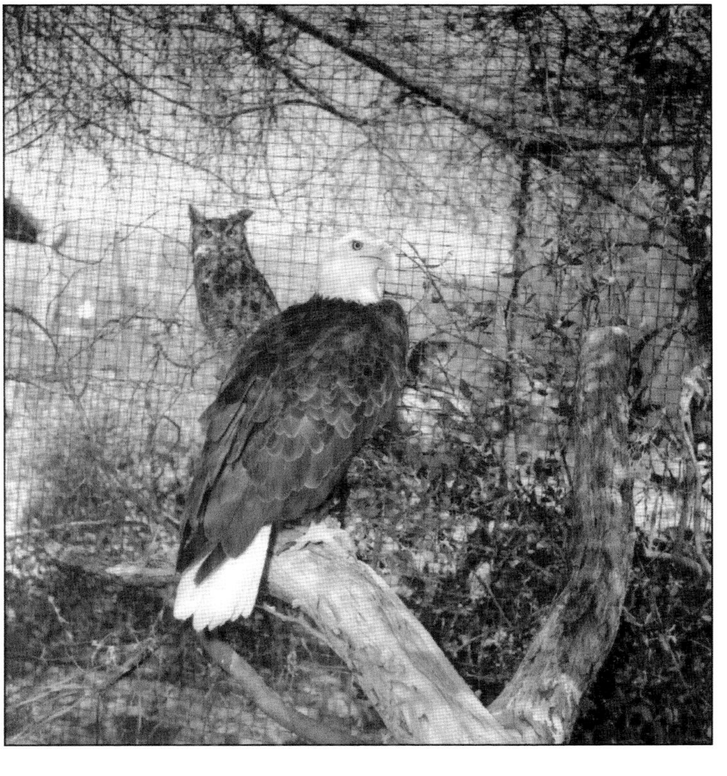

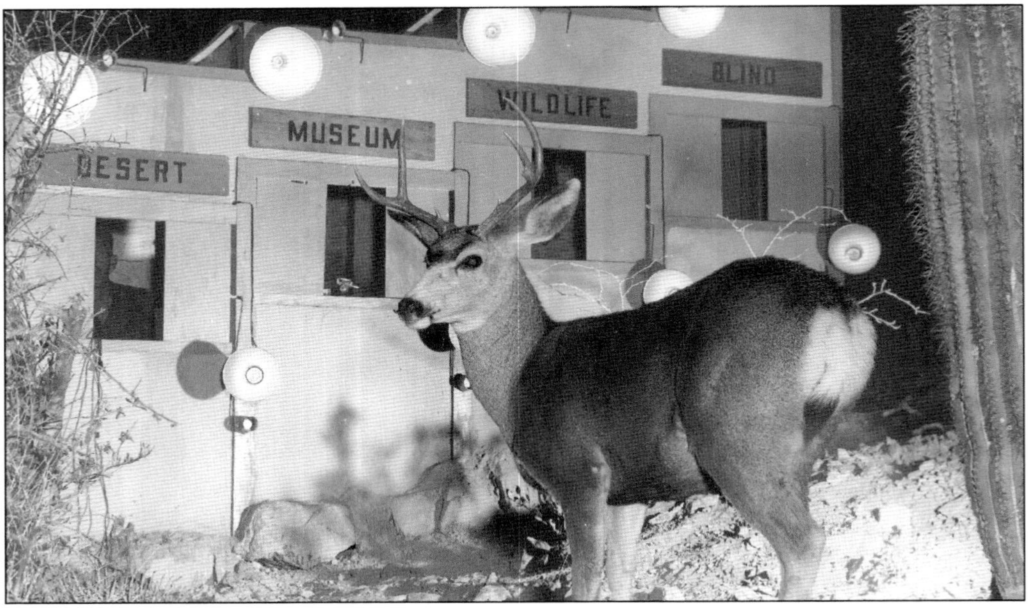

A slowly leaking water pipe in the corral at the Mountain House lured thirsty wildlife in the arid spring and early summer months. Lew Walker expanded the water hole and constructed the Desert Museum Wildlife photographic blind just above it. Lew produced wonderful photographs, many of which were published, including those in a special article about the blind that he wrote and illustrated for *National Geographic* magazine in 1958. (Photograph by Lewis Wayne Walker.)

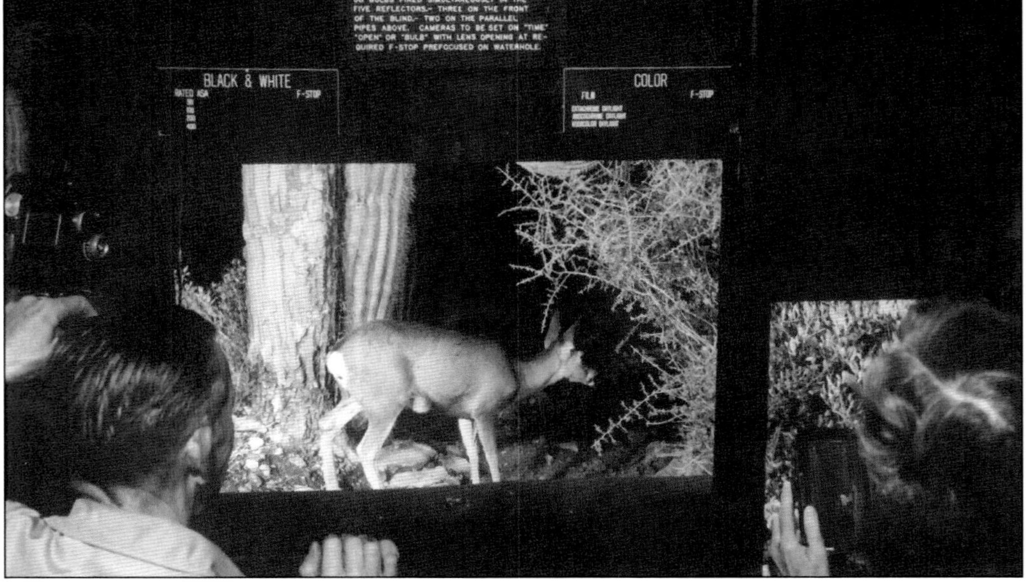

Several photographers could work inside the blind simultaneously and the floodlight activated by any one of them. The sign directs, "Cameras to be set on 'Time,' 'Open,' or 'Bulb' with lens opening at required f-stop, prefocused on waterhole." Reservations were required. Living at the museum, Walker oversaw the blind's use, and for several weeks in the early summer he was host to photographers from around the country.

In the early morning or evening, rabbits sometimes came to drink at the blind's waterhole, as well as a variety of birds, including doves, quail, thrashers, and wrens. Deer and javelina were the most frequent nighttime visitors. A large Sonoran Desert toad occasionally splashed into the water to soak or catch insects attracted to the blind's lights. Rattlesnakes visited at times. Some evenings, Lew placed raw meat near the blind to lure in skunks, bobcats, foxes, and other meat eaters. The blind provided an excellent site to observe animal behavior. Javelina babies came in walking under their mother's bellies and drank first, belligerently squealing if necessary to force the adults to wait. Deer showed little concern when sharing the site with rattlesnakes. The bulging eyes of a soaking Sonoran Desert toad protruding above the waterline spooked deer and javelina. And in one instance the photographers were also spooked when two fighting skunks entered the blind through a partially open door. (Both photographs by Lewis Wayne Walker.)

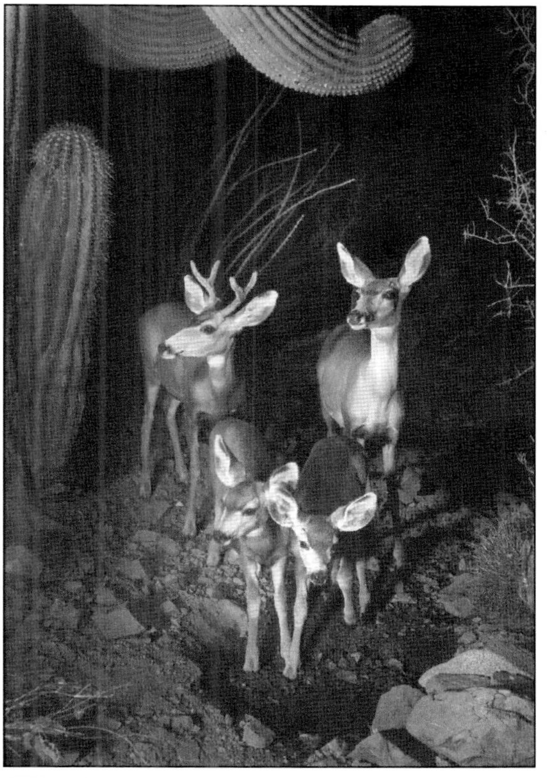

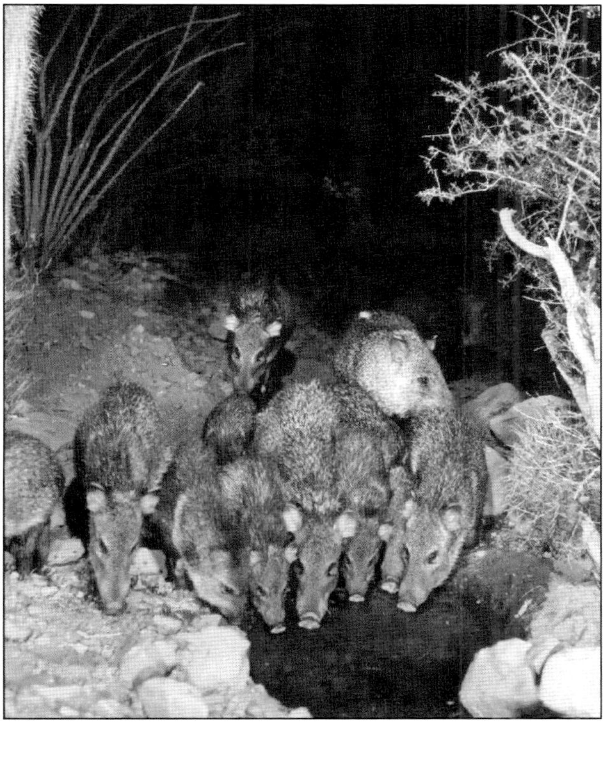

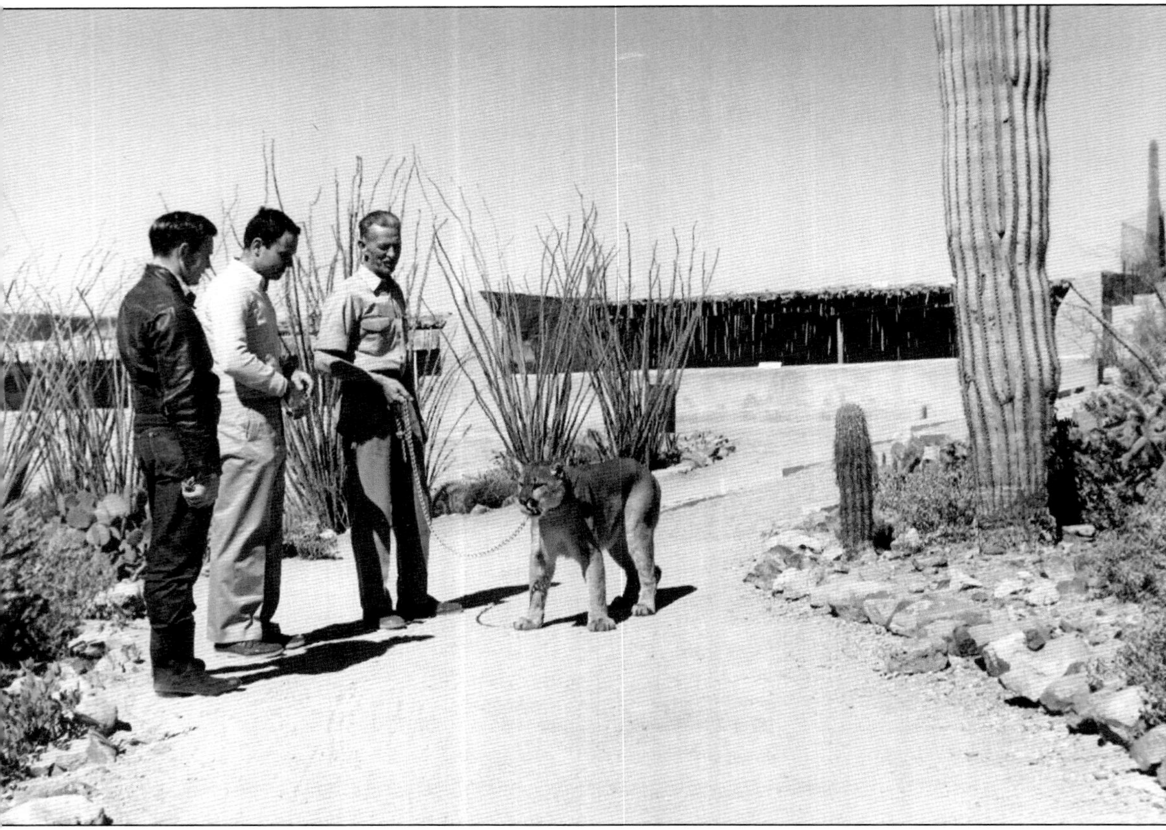

Skip Koth (left), Bill Woodin (center), and Lew Walker give George Mountain Lion II a morning walk on the museum grounds. The animal line appears in the background along with an abundance of saguaro cactus and ocotillo.

The wooden post was a mistake and was replaced by a metal one when it toppled. Additionally, it was thought that the caliche layer under the soil in the prairie dogs' open enclosure would deter the animals from burrowing out, but after a time, prairie dogs began to pop up around the museum grounds. The wanderers were retrieved, the exhibit renovated, and lots of cement solved that problem.

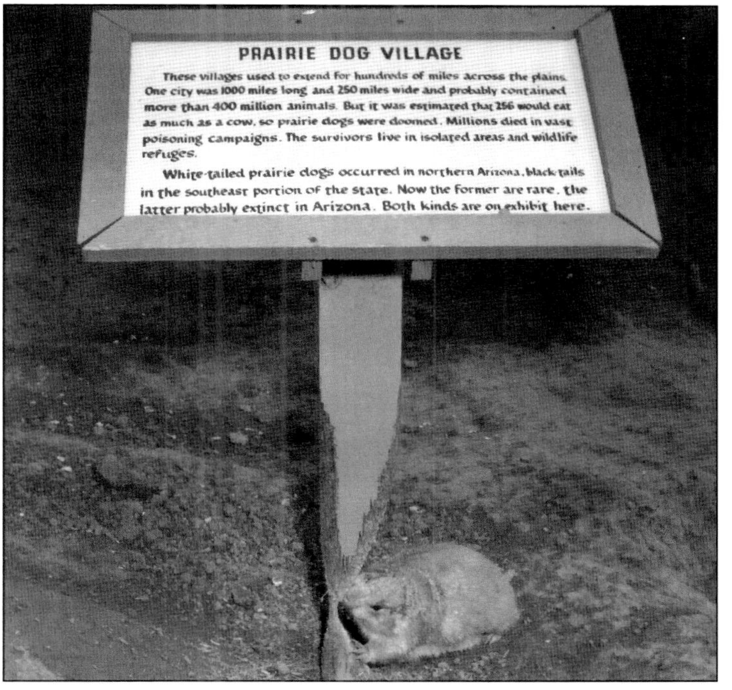

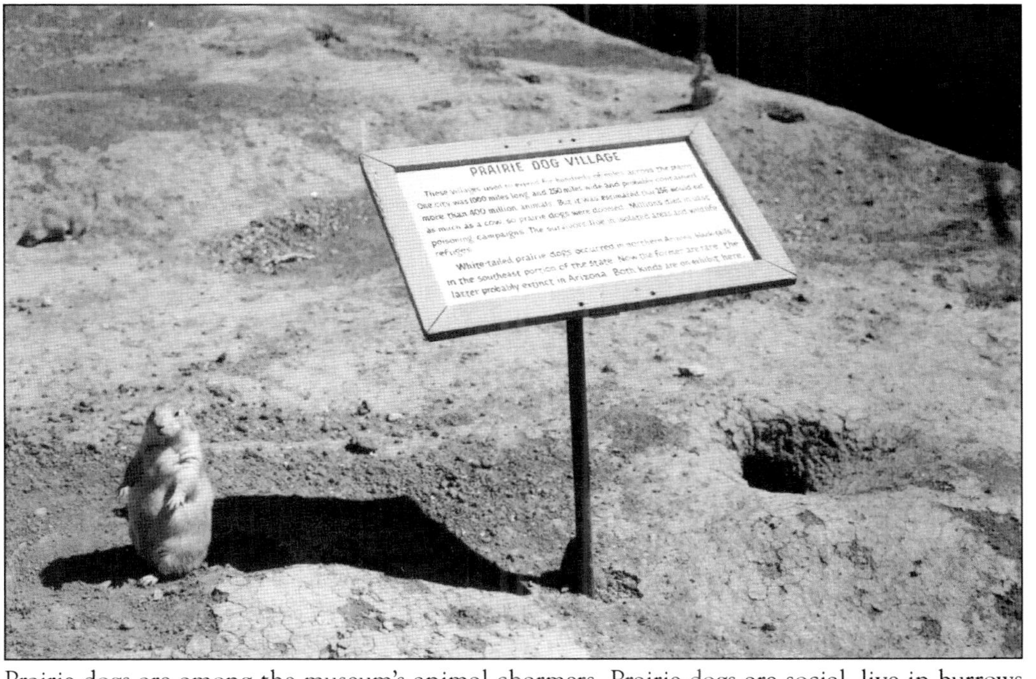

Prairie dogs are among the museum's animal charmers. Prairie dogs are social, live in burrows in extensive towns, primarily eat grass, and are not actually dogs but rather rodents. The "dog" name refers to their habit of giving bark-like warnings. When two prairie dogs meet, they often "kiss," meeting with open mouths, a means of determining if they belong to the same or a different family group. (Photograph by Wyatt Davis.)

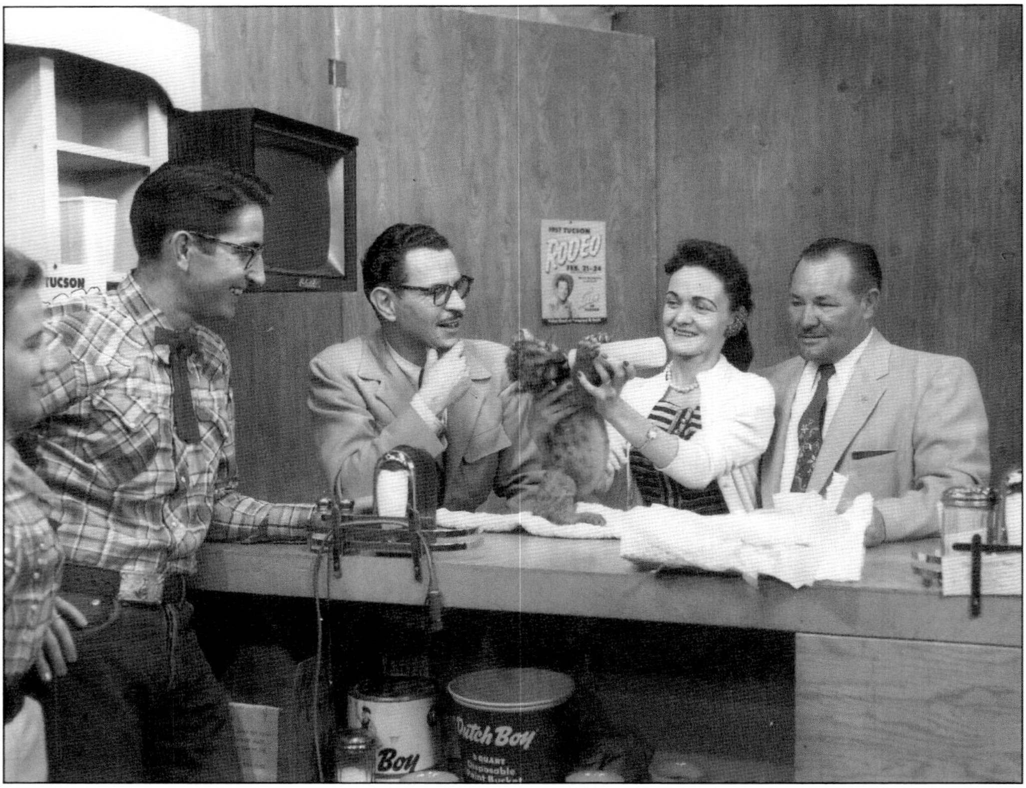

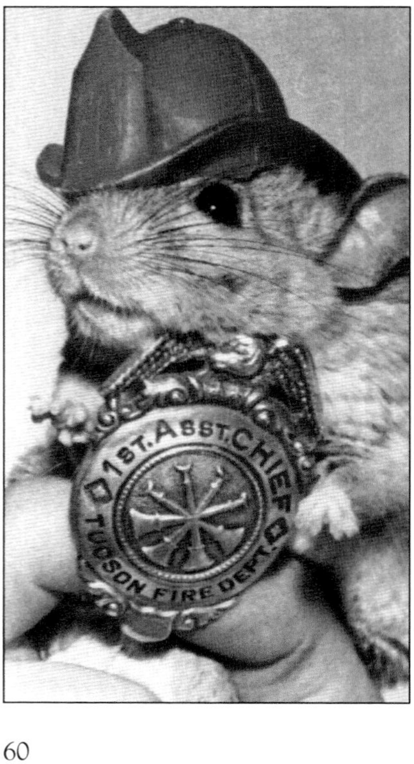

Hal Gras (third from left) was hired to produce and host the museum's weekly television show *Arizona Desert Trails*, the first of which aired in October 1953. Gras retained that position for three decades. Additionally, he took desert animals to schools and organizations to promote conservation of the desert biota. Gras was not a scientist, but in presenting the animals to children he taught the kids (and adults) about the desert and also about themselves. Woodrow Dent, a pack rat, helped Gras teach children about fire safety or wore a mortarboard to support Gras's admonition to children to gain a good education.

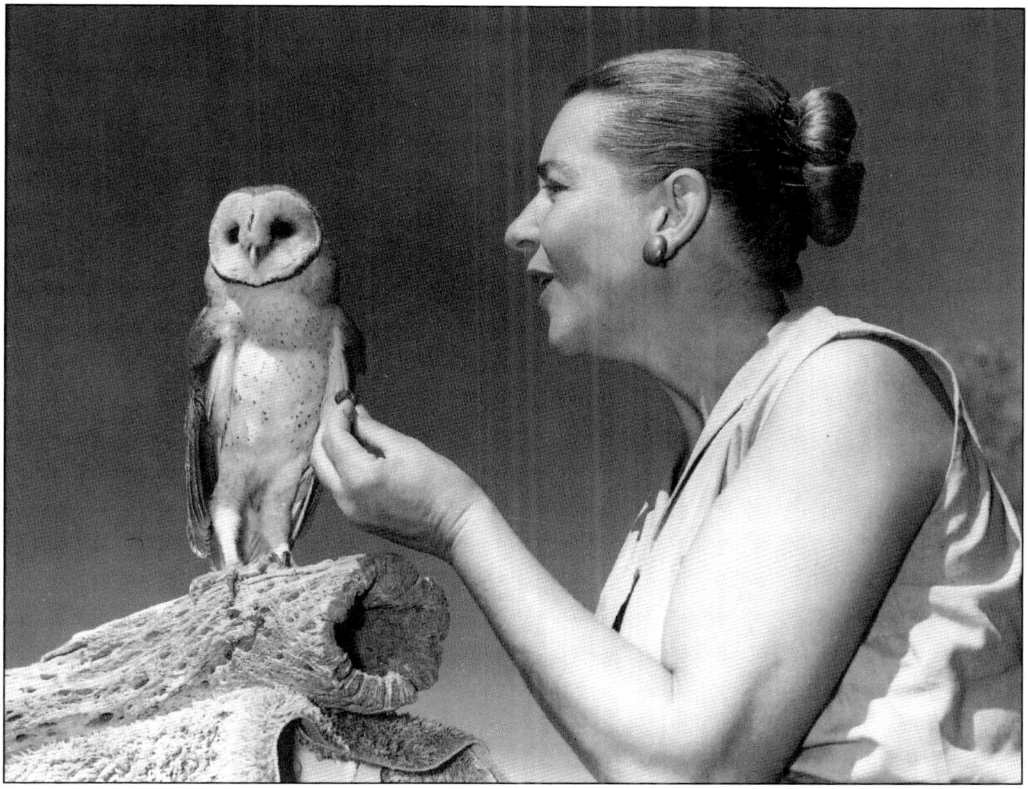

Renata "Natie" Gras provided loving care for the Ark animals living at the Gras's Tucson home, which soon became known as the museum's in-town nursery. There Natie received, raised, and rehabilitated orphaned or injured animals. Some of these could be released to the wild, but many could not. The latter became members of the Desert Ark or found homes at the museum or other institutions. At the annual members' meetings at the museum, Hal often reported on the number of animals residing in the Gras house and nursery. In April 1963, he reported that for that particular month there were 35 animals in the nursery, representing 27 species. In these photographs, Natie tempts a barn owl to eat and feeds a Harris ground squirrel from an eyedropper.

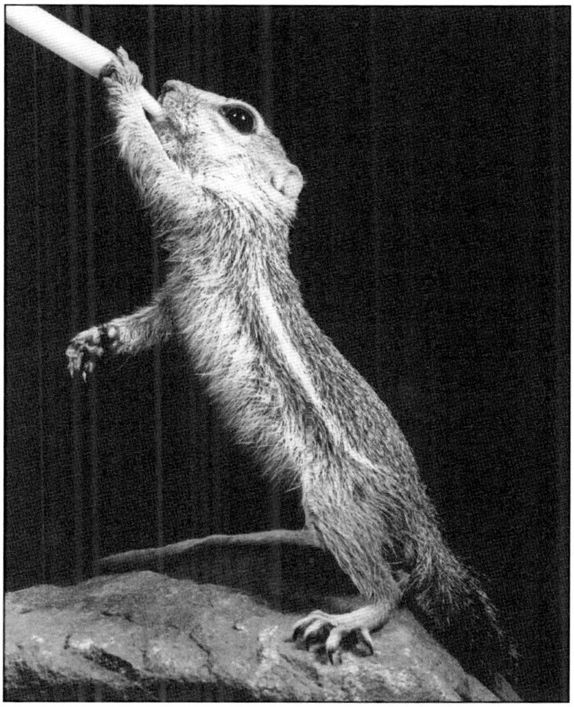

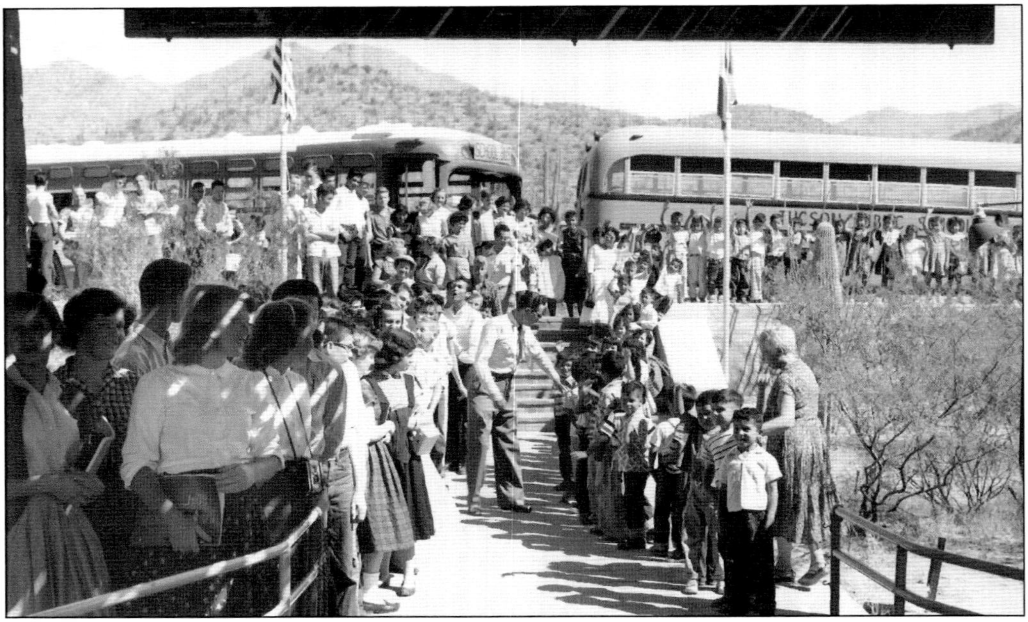

Education was, is, and always will be a major goal of the Desert Museum. Certainly the museum was achieving that goal on this particular day in 1955, when two school buses arrived at the same time and students of all ages lined up under Hal Gras's direction to tour the museum. (Photograph by Lewis Wayne Walker.)

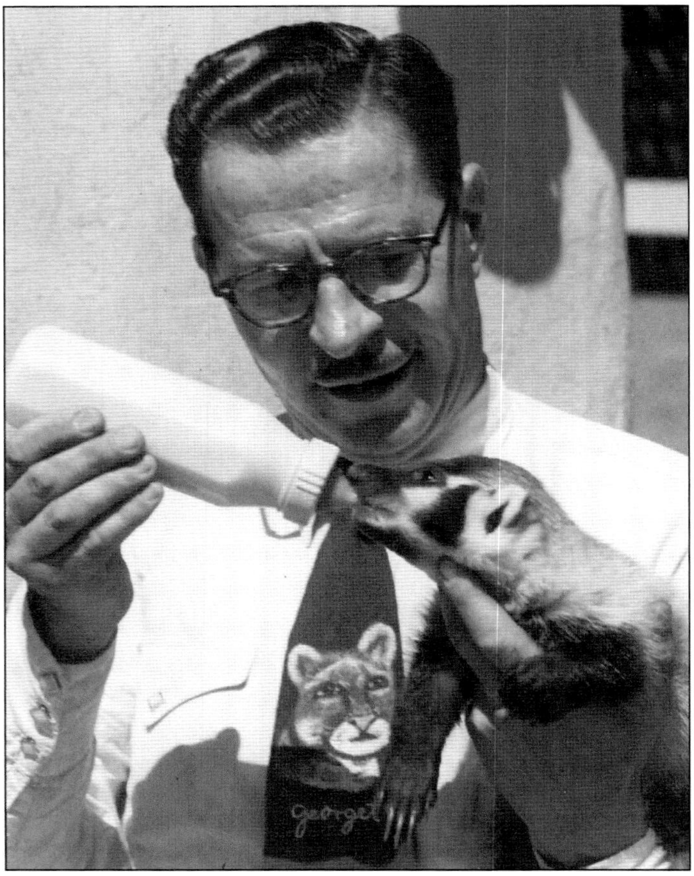

Over the years, Hal Gras introduced innumerable desert animals to thousands of people through personal appearances with the Desert Ark and on television. Baby animals were always popular with his audiences, and here he feeds a young badger. Hal was fond of George L. Mountainlion as an icon and museum spokesman and often wore a variety of ties adorned with George's image, these made by his wife, Natie. (Photograph by Wyatt Davis.)

Certain individual lions often stood in for the real George. Here Georgette peers from the Desert Ark above the drawing of George in 1956. An adult Jeanne Gras (Hal and Natie's daughter) stated that as a child she did not realize it was not normal for a family to have a very young lion resting in a baby playpen in the living room. Natie Gras reported that while confined there, the lion liked to watch the movement on the nearby television screen. (Photograph by Hiram L. Parent.)

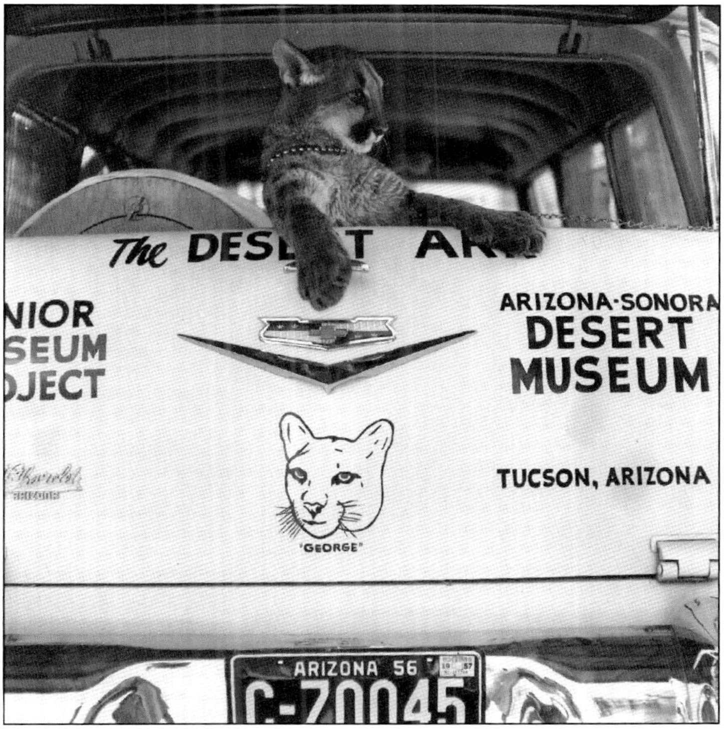

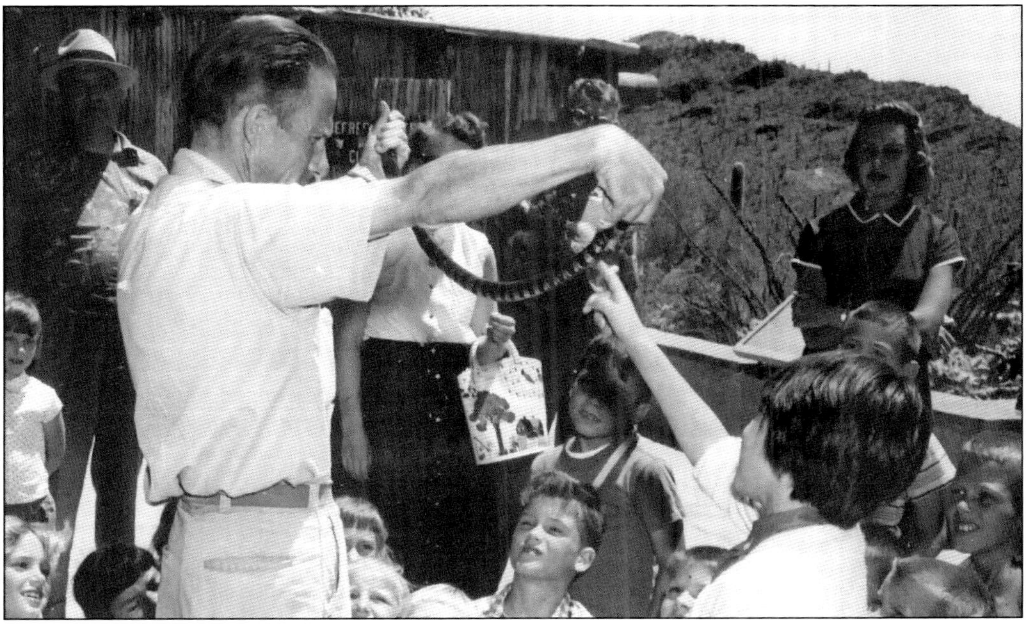

In remembering the establishment of the Desert Museum, Carr wrote, "One of our basic tenets, of course, was that school programs would be set up as a first order of business and that school use of the museum would be our continuing major concern." David Carswell, shown here talking to children about the king snake he holds, was an early employee charged to plan and implement formal educational endeavors at the museum. (Photograph by Wyatt Davis.)

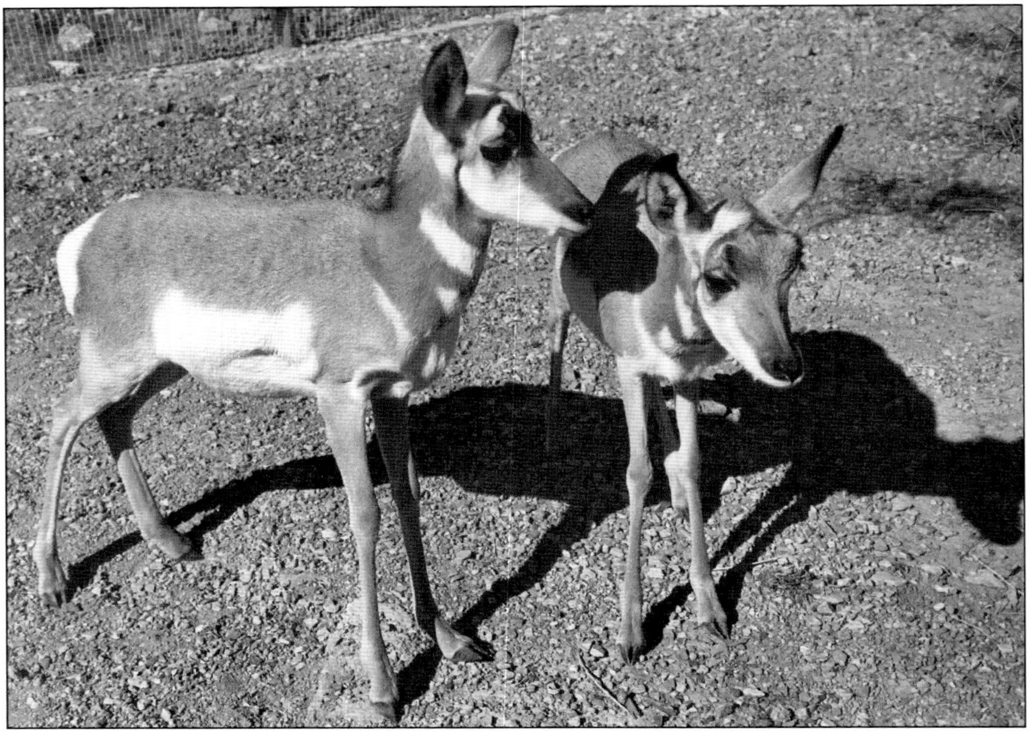

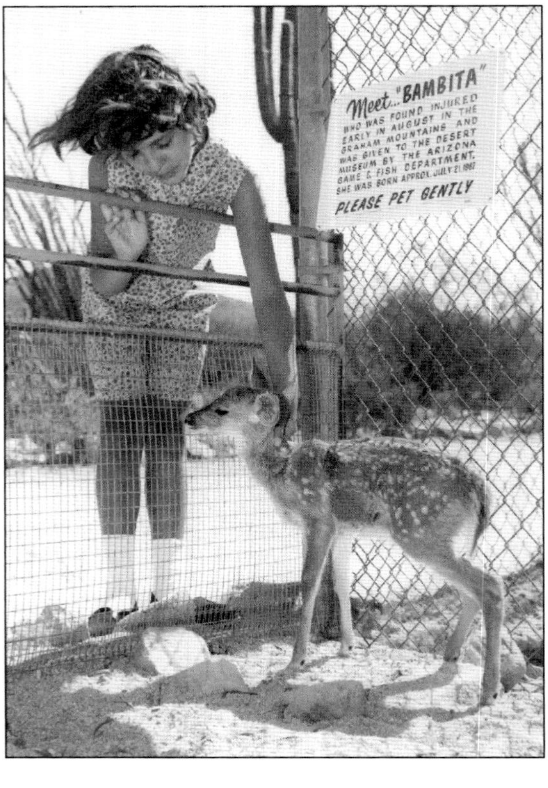

Early museum animals were often named. These pronghorn were Min and Gus (having come from the Mingus Mountains). In the wild, pronghorn prefer to live in open areas of little cover. They rely on their exceptionally good vision to spot predators from as much as 4 miles away and are the fastest runners in North America, reaching speeds up to 40 miles per hour. Both males and females have forked or pronged horns that are shed each year.

Also in early museum days, visitors, especially children, were allowed to touch cooperative animals in ways that would not hurt or disturb the animals or the youngsters. This girl reaches out to a cooperative fawn in a small enclosure along one of the museum's paths.

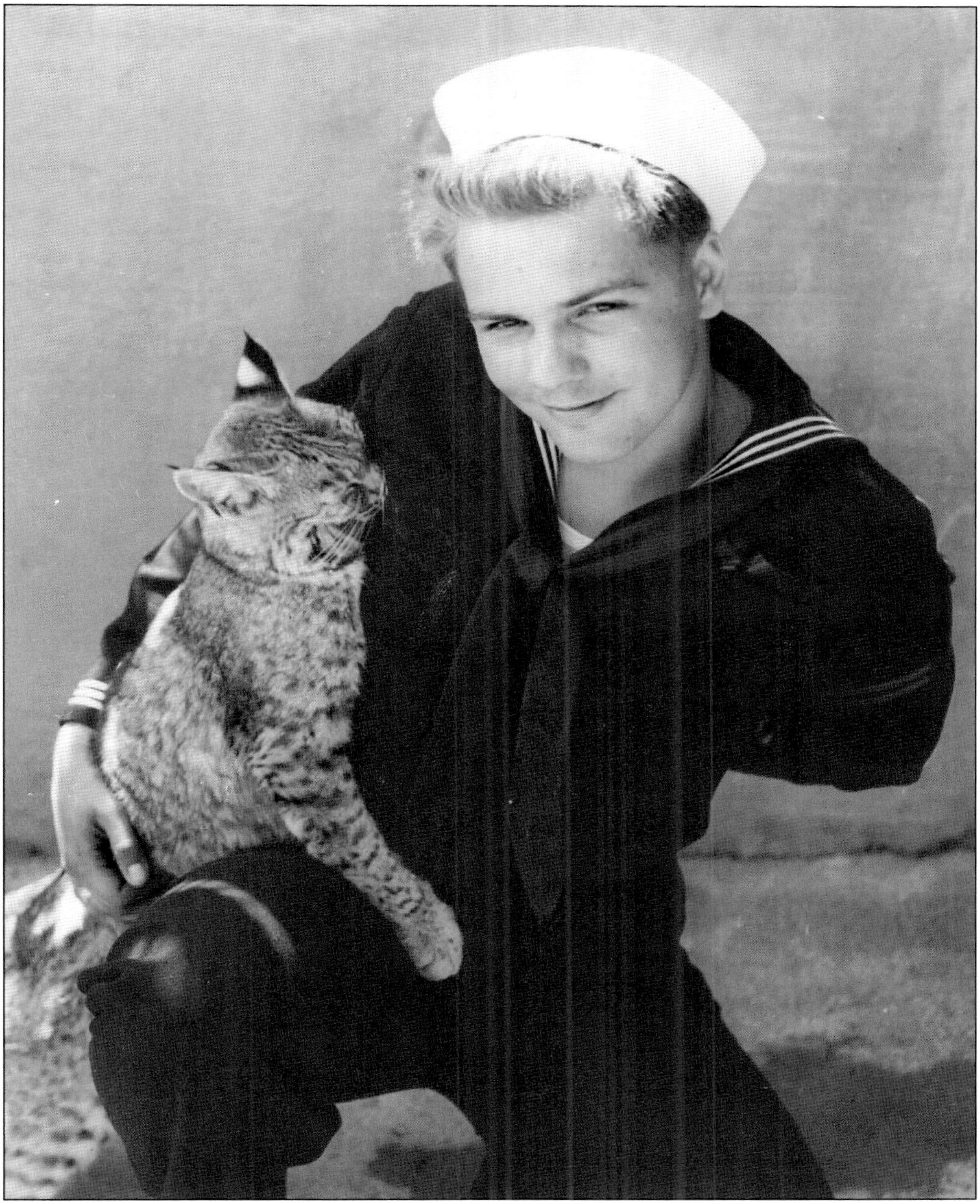

Sailor Bob Jones meets Diablo the bobcat. Diablo was a member of the Desert Ark and doubtless Hal Gras gave this serviceman an opportunity to be photographed with the cat. (Courtesy of Bob Jones.)

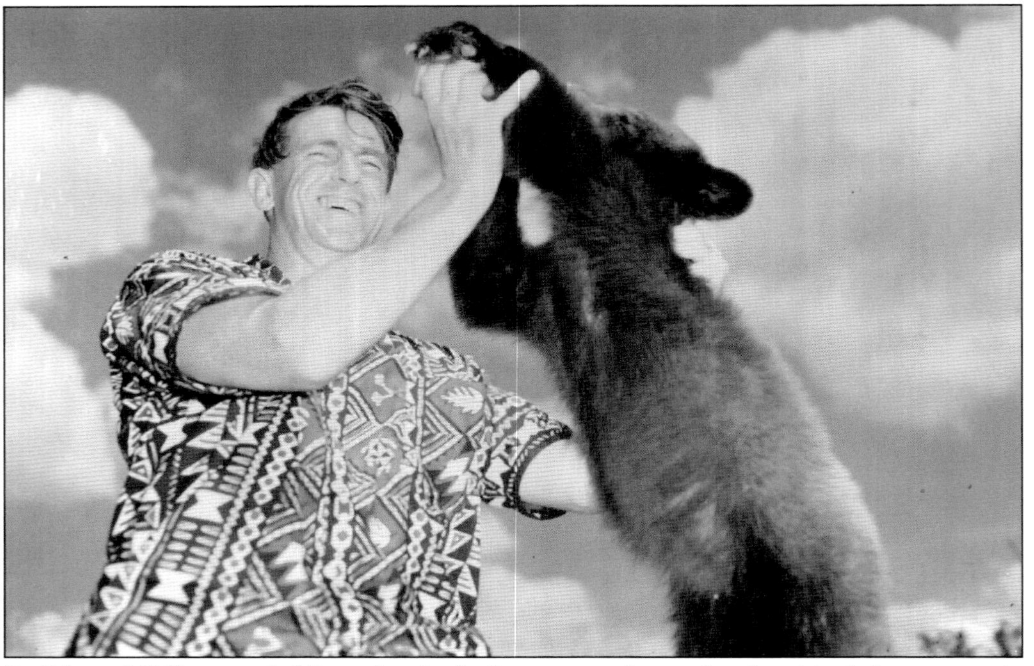

Sir Edmund Hillary was holding a bottle, feeding a young bear, when he visited the museum in the late 1950s. On May 29, 1953, he was the first person to ascend Mount Everest. Sir Hillary and his first wife, Lady Louise Mary Rose, were in Tucson visiting with their friends Watson and Lucy Smith. Watson Smith was a museum trustee who brought his guests to visit the museum.

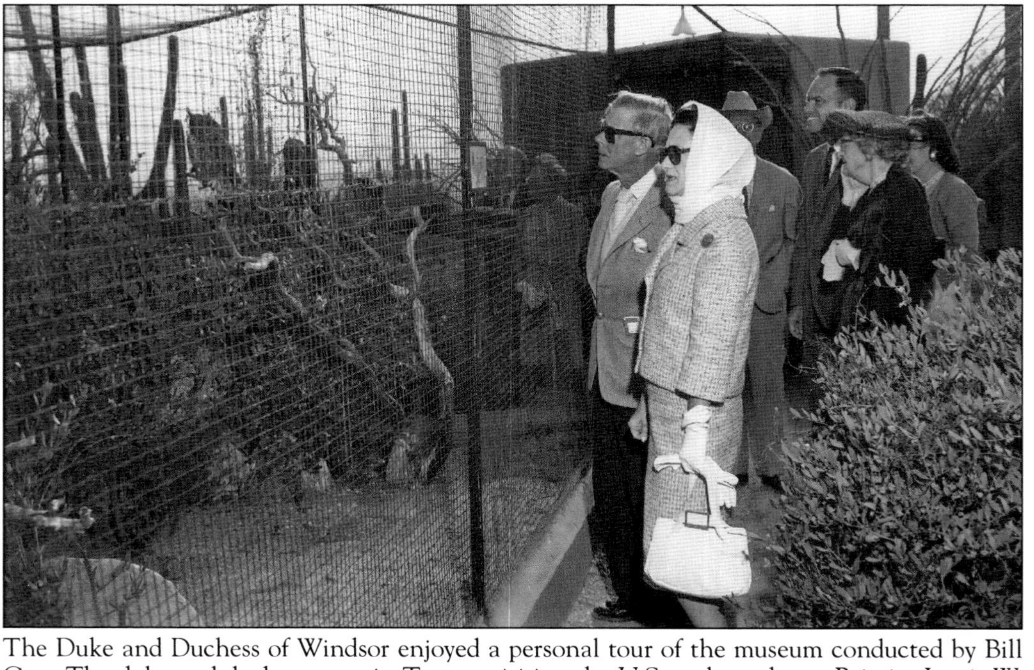

The Duke and Duchess of Windsor enjoyed a personal tour of the museum conducted by Bill Carr. The duke and duchess were in Tucson visiting the U.S. ambassador to Britain, Lewis W. Douglas. Peggy Douglas, hostess to the duke and duchess, was a museum trustee.

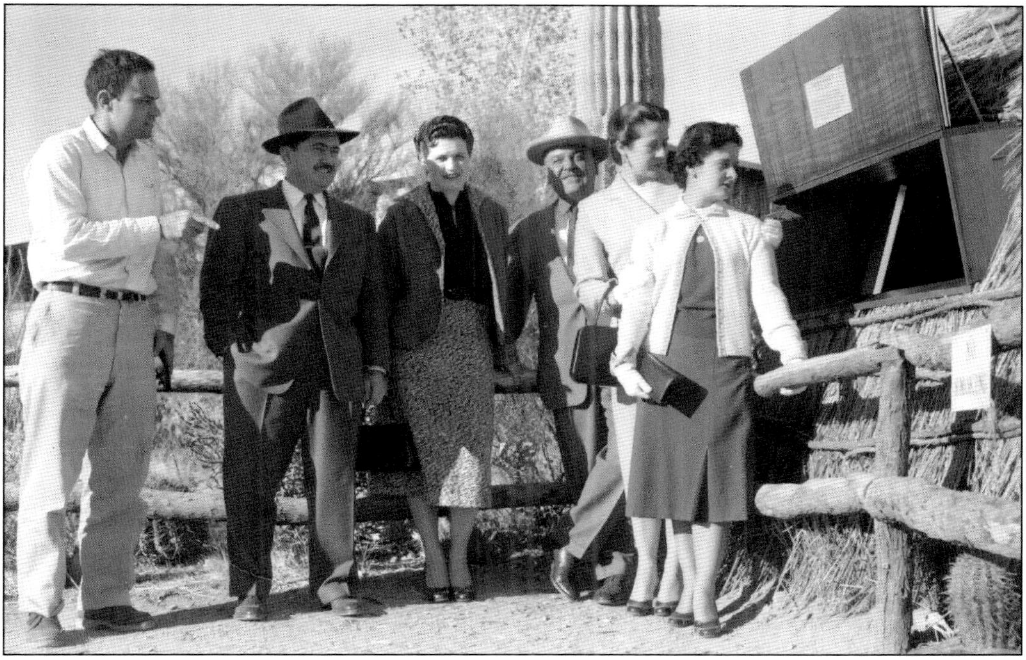

Bill Woodin (left) escorts museum trustee Alex Jacome and his family on grounds. The observation window of the shaish-ki, which allowed viewing of the Papago (Tohono O'odham) sand painting, appears on the right.

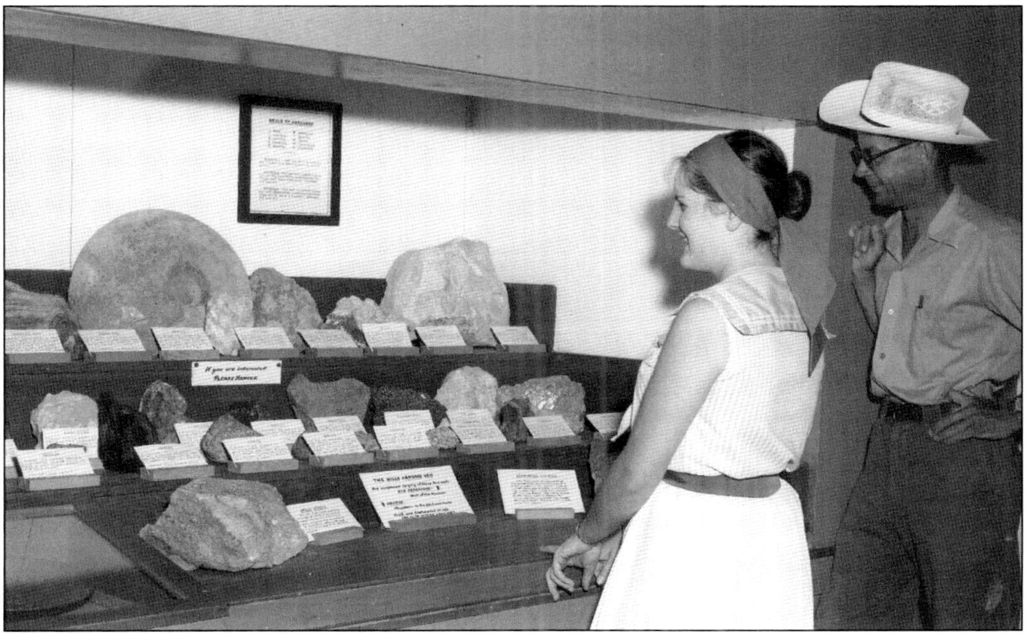

A long case containing 124 minerals was one of the earliest attempts by the museum to provide a public mineral display. Many of these minerals and the vast collection that has been obtained since owe much of their provenance to museum trustee J. Rukin Jelks Jr. (Photograph by Wyatt Davis.)

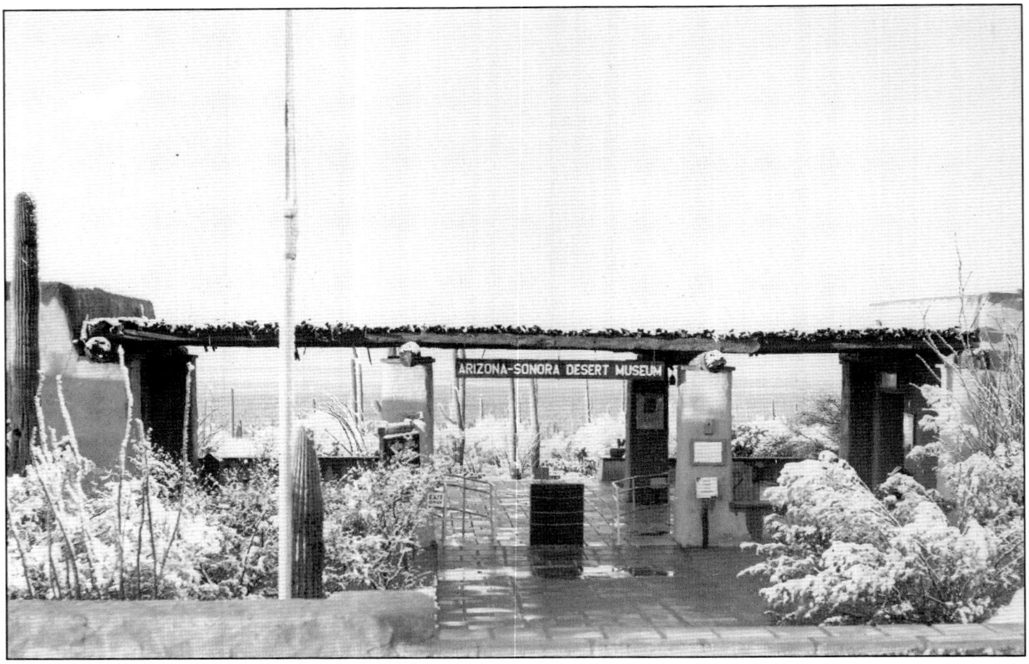

Occasionally the museum receives snow. In November 1958, a storm blanketed the museum with seven inches of it. The result was a scenic wonder, a collapsed bird enclosure roof, and a total of only three intrepid museum visitors.

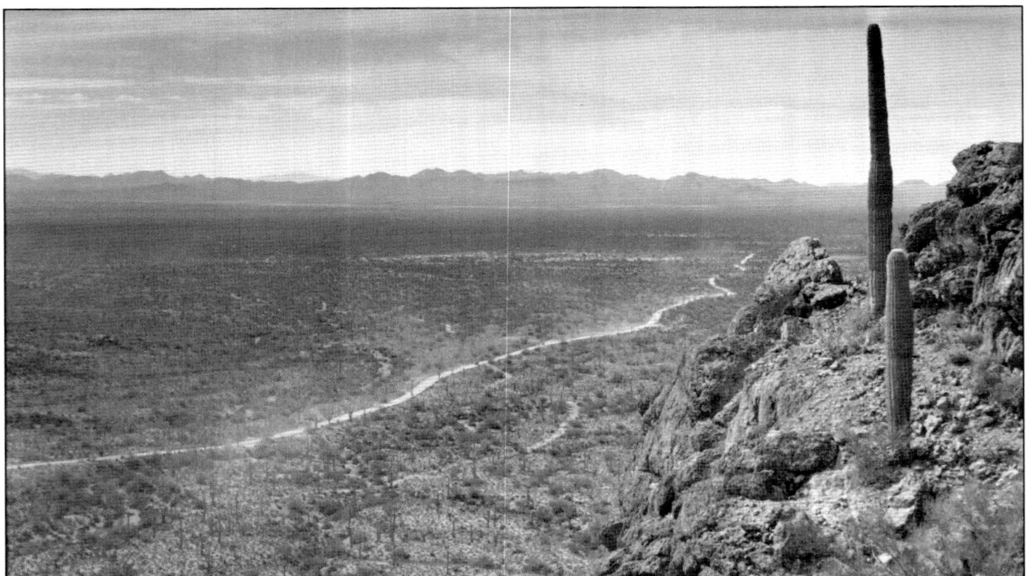

A look at Gates Pass Road and Avra Valley from the vantage point of Gates Pass in the 1950s reveals a dirt road leading toward Kinney Road. Until 1961, the road to the Arizona-Sonora Desert Museum remained unpaved. In the early years, the museum had no telephone because there were no phone lines. A radio phone was used for communication with Tucson; however, its reception was poor because of the Tucson Mountains.

Two

1960s

In December 1954, Bill Carr resigned as director due to poor health. The museum was experiencing financial difficulties. There had never been an admission charge, and the initial Charles Lathrop Pack Forestry Foundation funds were largely expended. It was time to regroup. An admission charge of 50¢ was instituted. The staff of nine went to five and their salaries were reduced. William H. "Bill" Woodin was elevated to director. Expenses were cut to the bone. Yet despite some problems, much had been accomplished in the first eight years.

Board members and many community companies and individuals were generous with time and money. Pack money built the Tunnel and Water Street. Another board member's donation funded the Amphibian Room. And the staff, whose numbers gradually increased, were hardworking and dedicated to the museum and its purpose.

In the 1960s, basic physical background problems were remedied. New restrooms, shade ramadas, and a new sewer system, water supply and storage, animal kitchen, administration wing, remodeled indoor-outdoor restaurant facility, and a workshop building were built to support the increase in visitors and exhibits. In 1961, Gates Pass Road was paved, and in 1963, the museum finally had a telephone line.

New exhibits included a coatimundi enclosure, walk-in tortoise enclosure, phase one of the Demonstration Desert Garden, vampire bat cave, and otter exhibit. By 1966, an animal inventory at the museum counted 876 residents. Births of animals were numerous and the young popular with both staff and visitors. In 1969, these babies included six Mexican gray wolves, thick-billed parrots, and two kit foxes. A recently obtained six-month-old female jaguar joined them.

The museum's exhibit techniques were being copied by other institutions, and Desert Museum exhibit staff often served as consultants or exhibit builders for them. The museum had been instrumental in the declaration by the Mexican government making Isla Rasa in the Gulf of California a protected refuge for nesting birds.

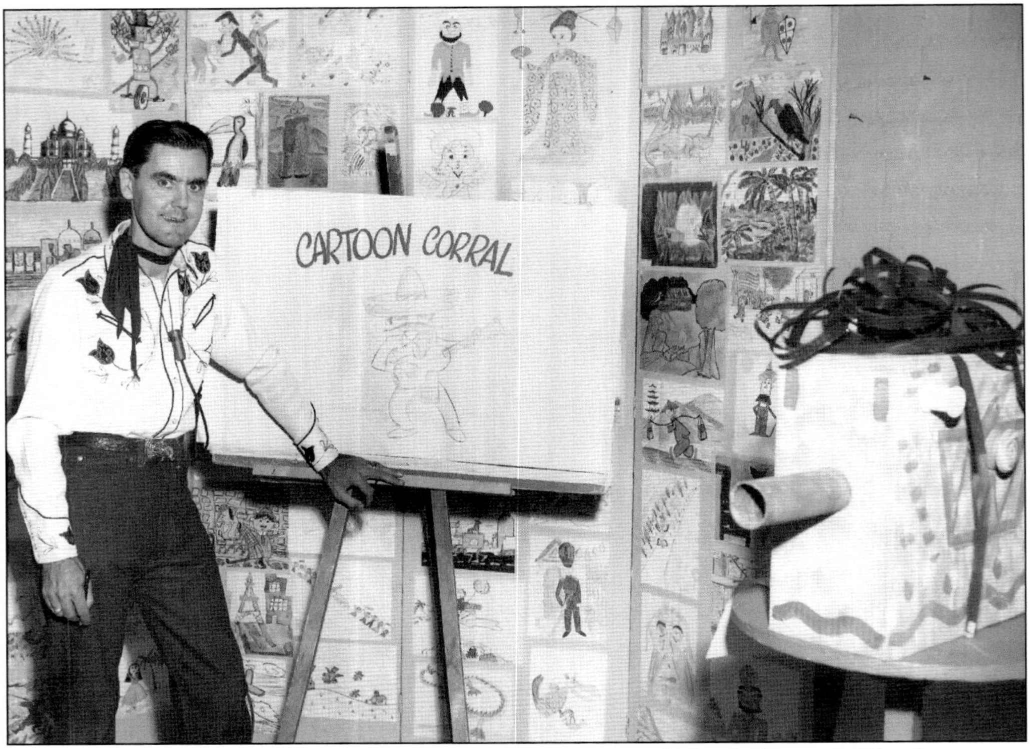

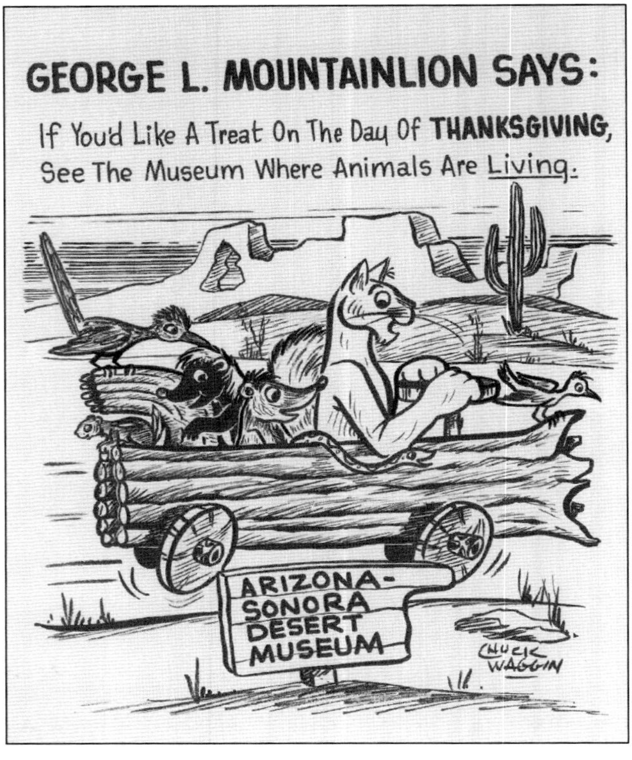

"Chuck Waggin" (Charles Amesbury) was art director for a local Tucson television station and appeared weekdays on the children's program *Cartoon Corral*. He drew many humorous illustrations that appeared in the *Arizona Daily Star* for the museum's George L. Mountainlion column. For several years, his wife (as George L. Mountainlion) wrote short rhymes that accompanied the drawings in the column. In his newspaper column, George L. Mountainlion wrote about happenings at the museum, provided natural history information, and often urged his readers to visit the museum. Chuck Waggin's drawings provided all his animal characters with a great deal of personality. (Both photographs by Ted Offret.)

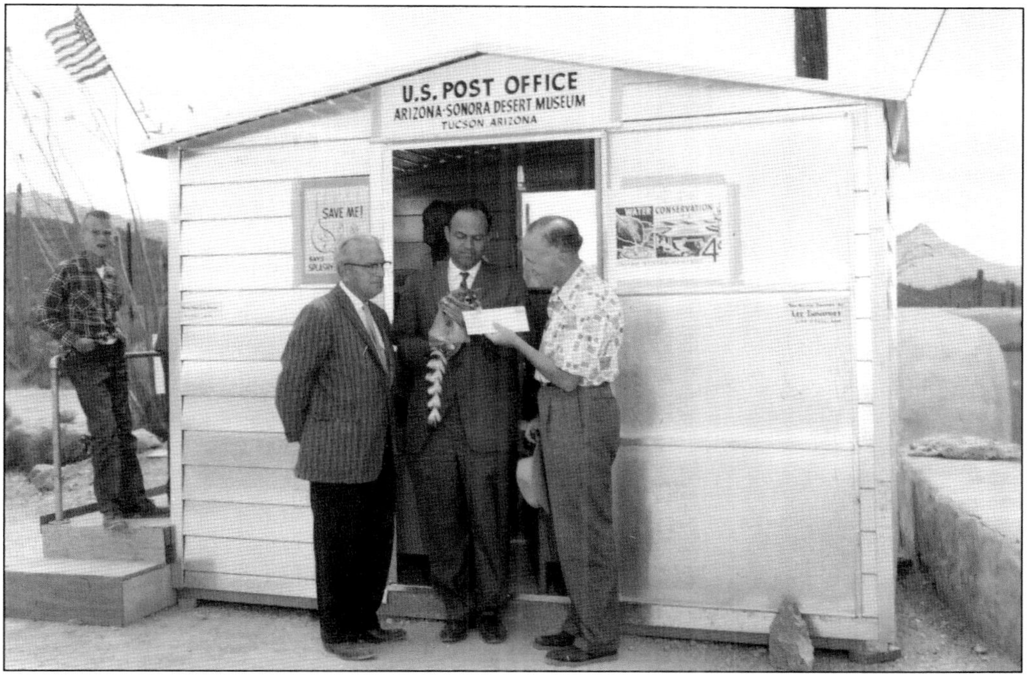

In 1960, the U.S. Post Office issued a new 4¢ water conservation stamp. The stamp was issued to stress the conservation "of our most precious and most limited natural resource—water." The Arizona-Sonora Desert Museum received national recognition on Saturday, April 23, when the Post Office Department established a one-day postal substation at the museum entrance. Philatelists from every state in the union and from all corners of the world forwarded letters and cards to Tucson to receive a rare Desert Museum Watershed Exposition cancellation and a special rubber stamp cachet in honor of the observance. Below, Bill Carr (center) is pointing out the importance of soil conservation to a visitor as director Bill Woodin looks on.

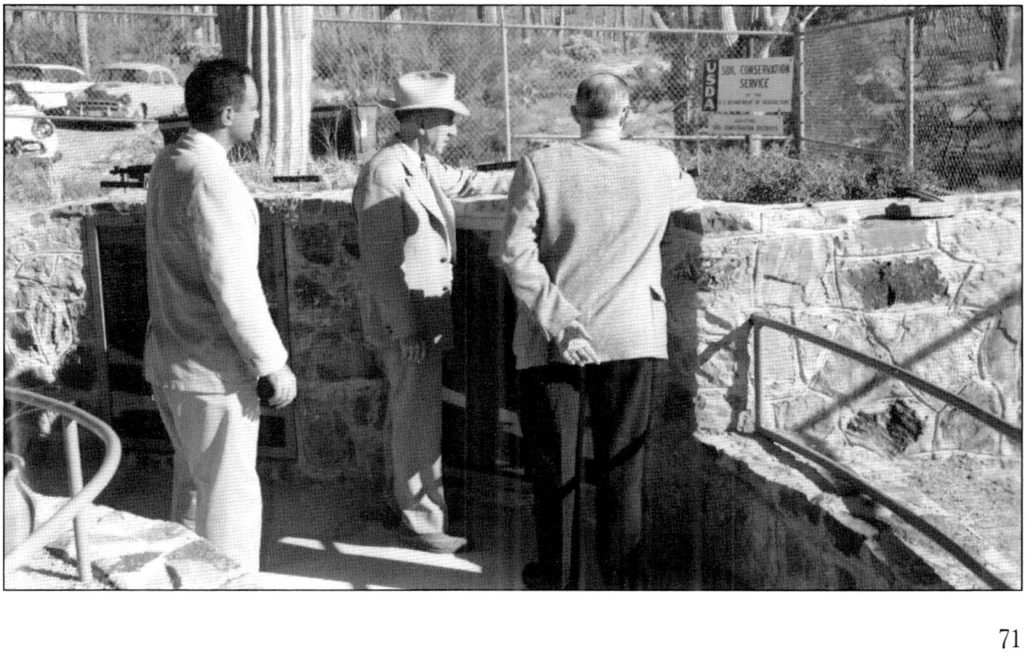

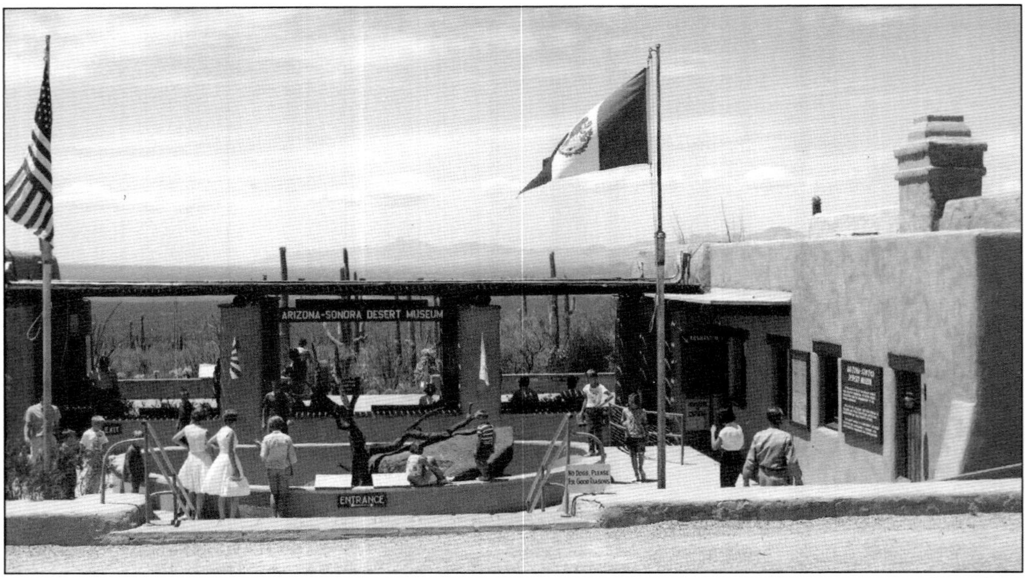

The Sonoran Desert lies in both the United States and Mexico, and museum personnel work in both countries; therefore, the museum flies the Mexican flag as well as that of the United States. The museum's entrance has changed over time. This view is from the early 1960s. The structure at the center held live lizards. (Photograph by Mervin W. Larson.)

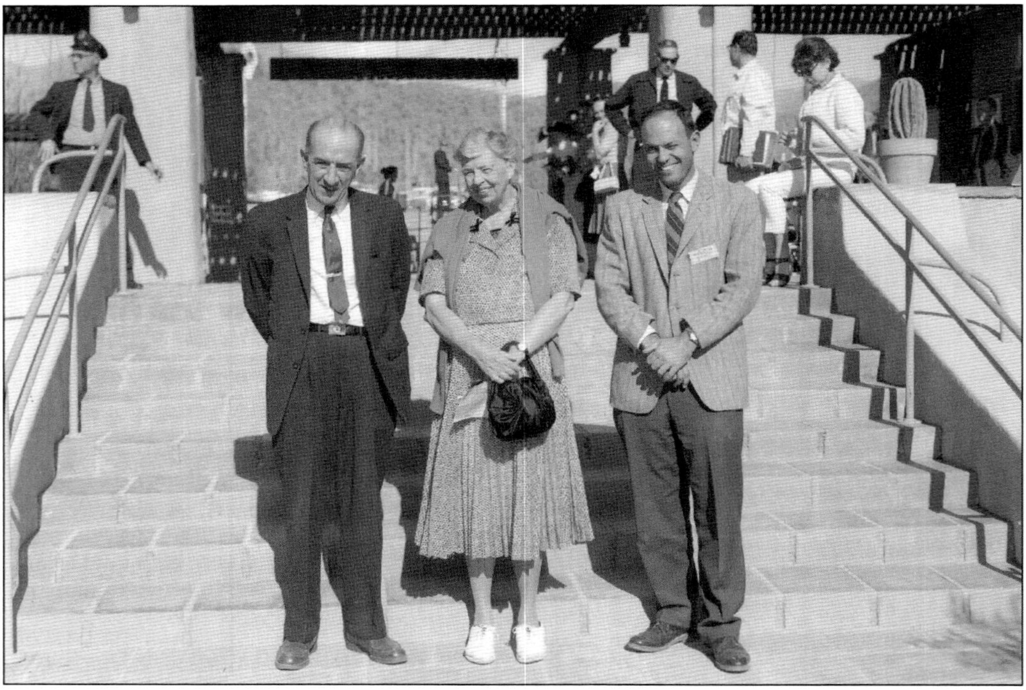

Eleanor Roosevelt is flanked by Bill Carr (left) and Bill Woodin on her visit to the museum in 1961. She reminisced with Carr about Bear Mountain Museum, which she had visited when he was director there. Bill Woodin's grandfather served as her husband's secretary of treasury during the financial crisis of 1933.

Associate director Lew Walker explored the Gulf of California islands starting in the 1930s. He was an early and successful advocate for Mexican legislation that preserved the nesting populations of more than one million Heermann's gulls and royal and elegant terns on the less-than-1-square-mile Isla Rasa.

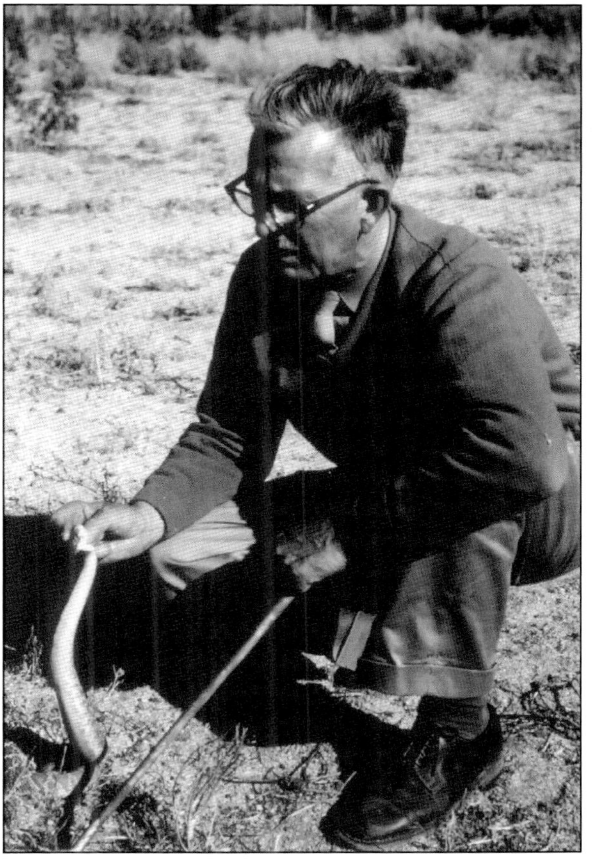

Through the efforts of the Desert Museum, particularly those of Lew Walker, the Audubon Society, and Mexican wildlife officials, Isla Rasa in the Gulf of California was declared a waterfowl sanctuary in 1964. Rasa was the first sanctuary in the Gulf and served as an impetus for protection of all the Gulf islands by the Mexican government in 1978. Terns incubate their eggs in this picture. (Photograph by Lewis Wayne Walker.)

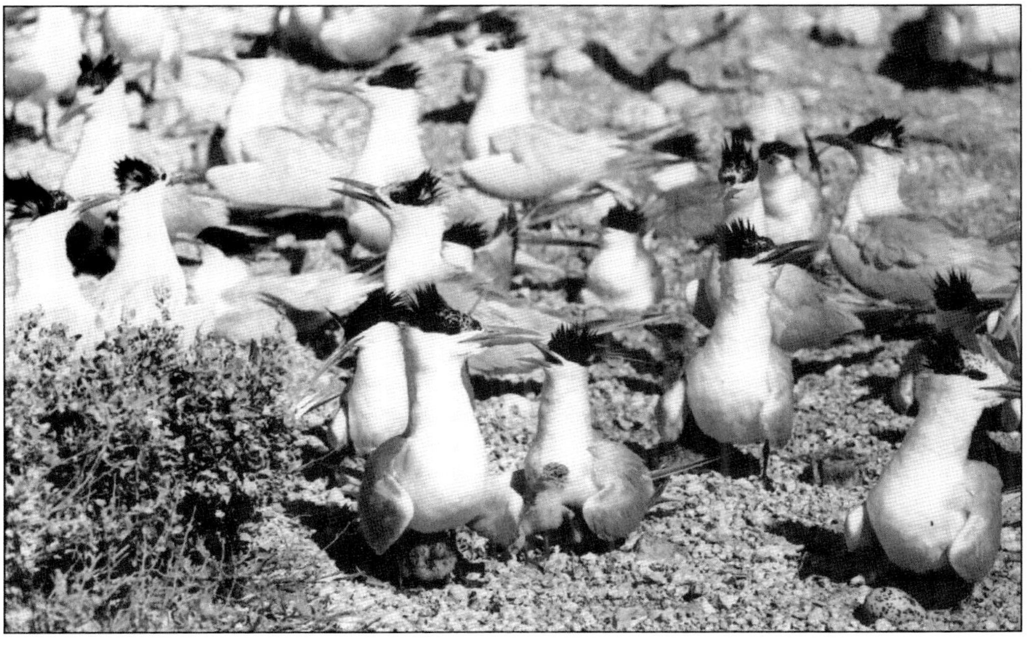

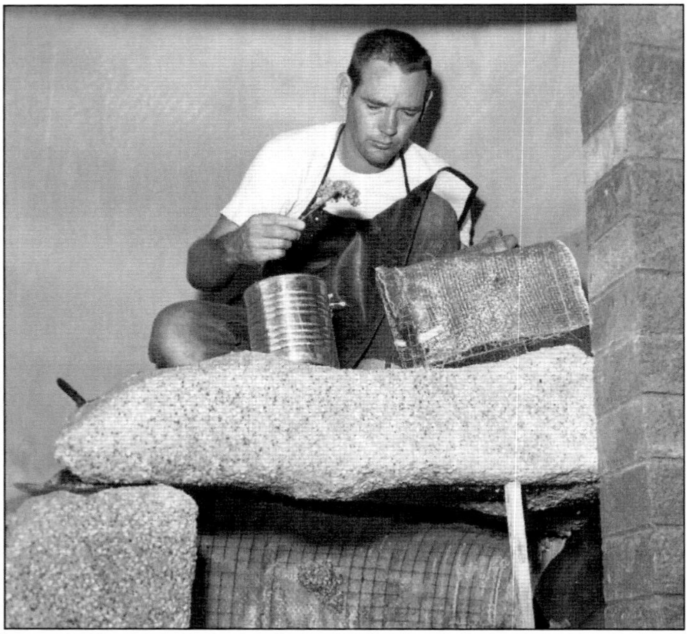

Merv Larson is shown here working diligently on artificial rocks. His expertise in the field of exhibit design and construction of artificial rockwork became renowned worldwide as the Arizona-Sonora Desert Museum grew during the 1960s. In the 1960s, artificial rockwork was created using air pressure to shoot concrete and plaster onto metal rods and laths, which was later sculpted and painted by hand to resemble a natural formation. In the photograph, Larson is constructing artificial rockwork using polyester resins and crushed rock.

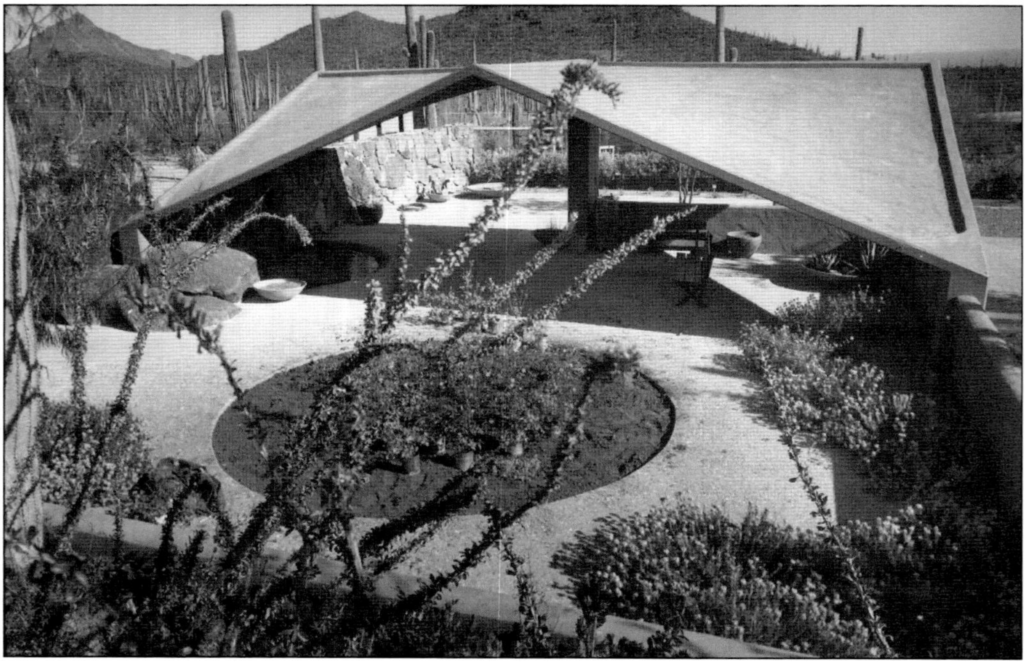

The Demonstration Desert Garden opened in phases beginning in 1963 and continuing through 1971. Its creation was initiated by the Arizona-Sonora Desert Museum, *Sunset Magazine*, and the local community, including assistance from the University of Arizona and business groups. The purpose of the garden is to provide aesthetic enhancement of the grounds, as well as to be a center for experimentation and exhibit of new and varied ideas for desert landscaping. The vegetation in the garden includes plant species indigenous to the Sonoran Desert in Arizona and Sonora. (Photograph by Peter Balestrero.)

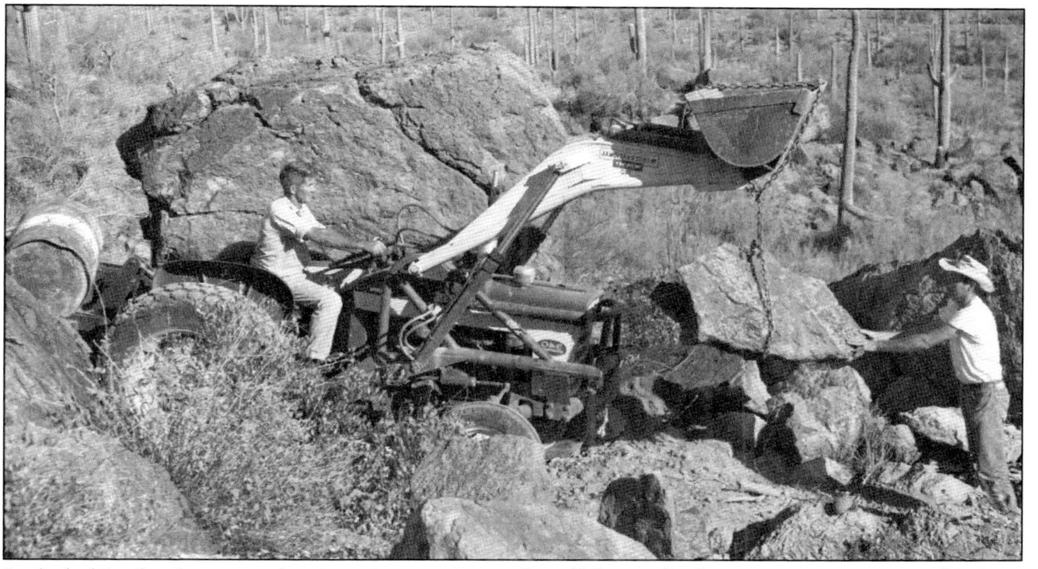

Included in the desert garden is an impressive rock wall from which water runs into a pool below. This portion of the garden is often a focal point around which special museum or personal events are held, including weddings, receptions, and memorials. Gathering the boulders for the construction in the 1960s are Cletus Hogan (on the tractor) and Merv Larson.

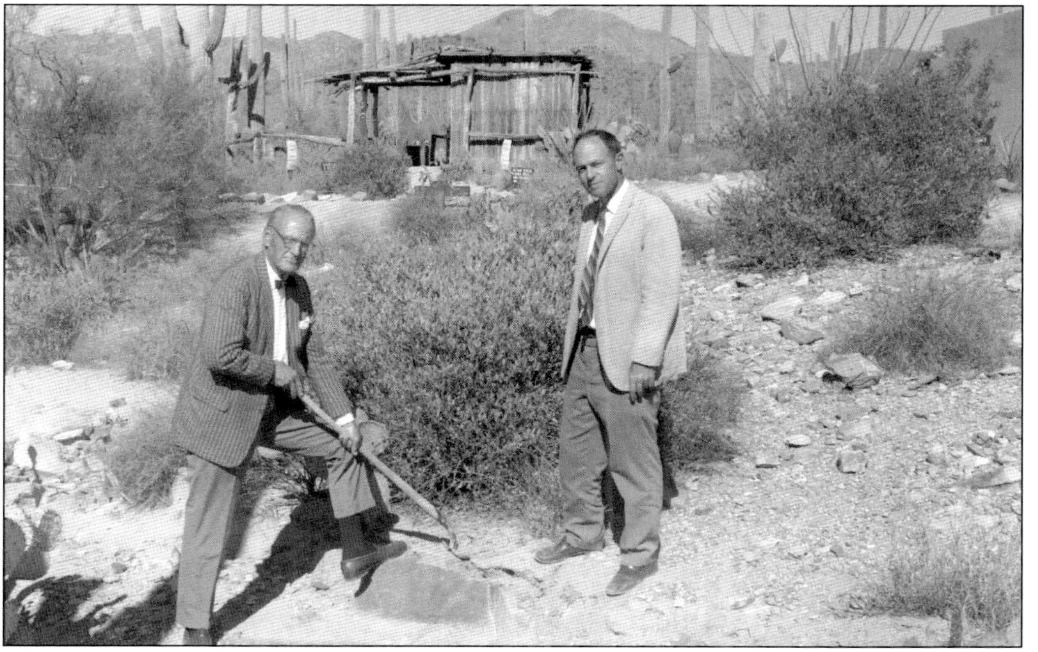

The ground-breaking for the beaver-otter-sheep exhibit was held January 3, 1966. John Jameson (left) and Bill Woodin appear eager to initiate construction of this riparian area. Jameson was then president of the museum's board of trustees. Later, as a museum employee, he served as development officer and then as chief administrative officer. Jameson was noted for his fondness for otters, and his extended family donated a pair of otters to the museum in his honor on his 70th birthday.

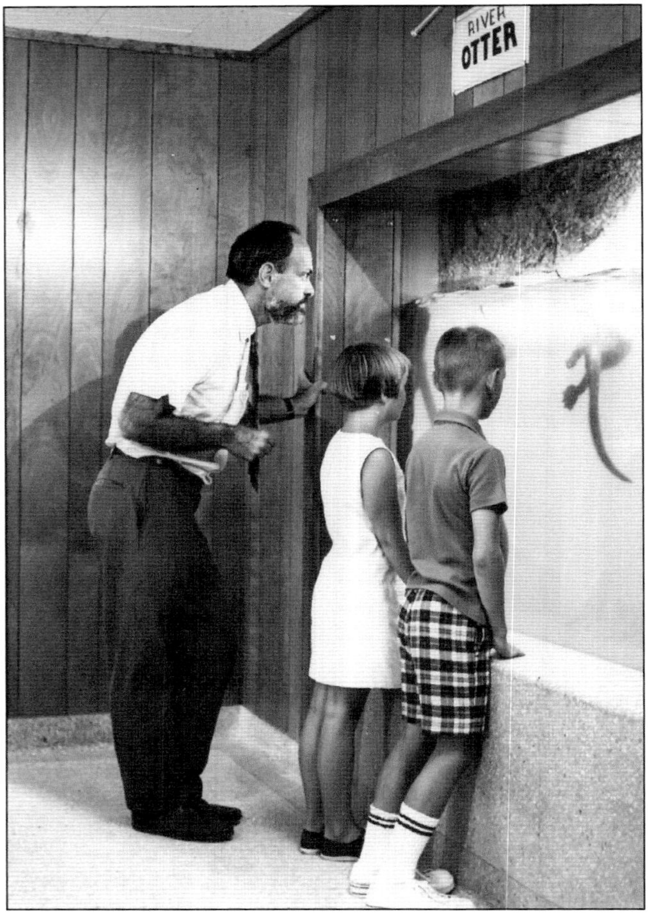

Director Bill Woodin and two young spectators watch a river otter perform in the recently built exhibit. Visitors could observe the otters on two levels as the animals swam on the surface or underwater. The exhibit included a special den where visitors could view the otters as they rested.

An early display of historical artifacts associated with beavers and their important role in American history supplemented the live beaver exhibit. Bill Carr was especially fond of beavers. He had collected books and memorabilia about them for many years and provided the items for the exhibit. (Photograph by Dan Sheehan.)

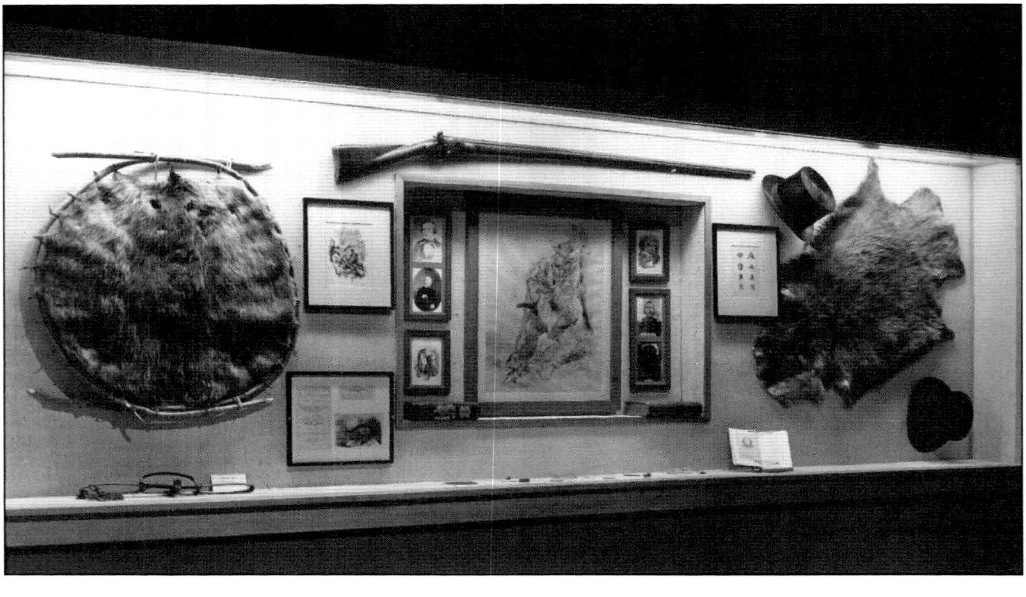

A popular beaver model was once displayed as an accessory to the beaver-otter-sheep exhibit. Museum exhibit techniques called a dissolving unit were used in this exhibit. Visitors could activate the exhibit, causing the beaver model to be transformed from a full body model to fade into that of its skeletal structure, and then slowly revert back to the body model.

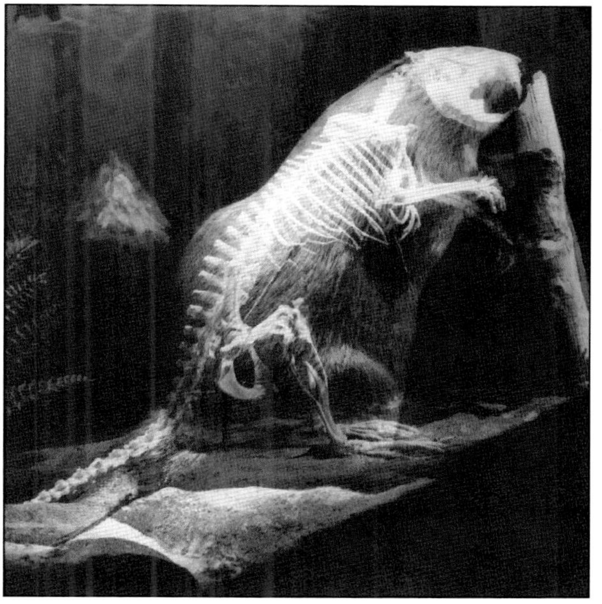

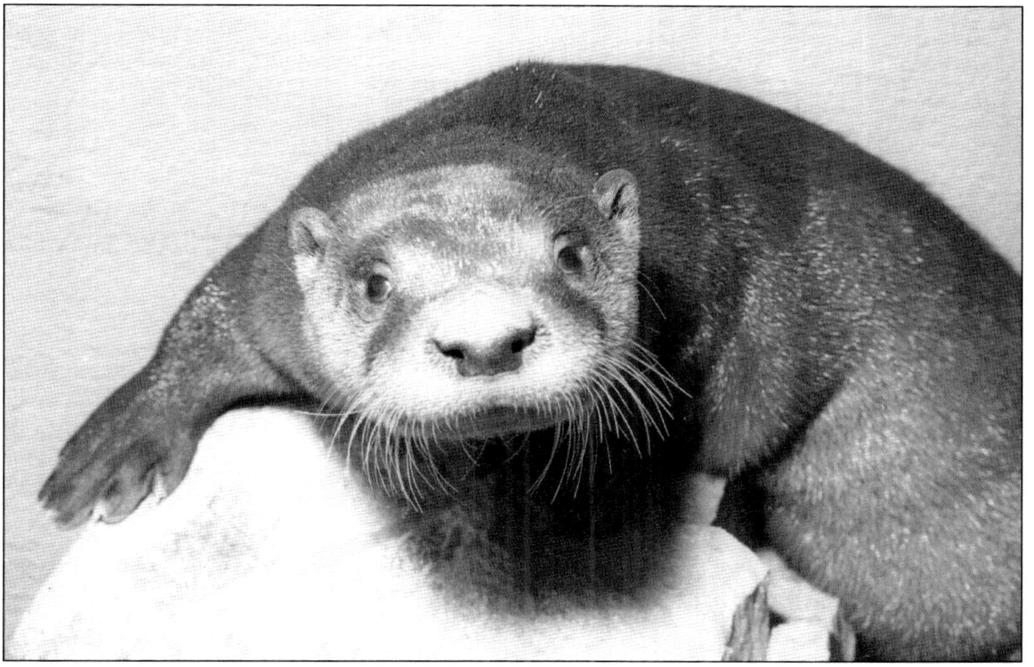

The otter exhibit was designed to simulate a riparian area once common along the banks of the Santa Cruz River near Tucson. Zoo officials worldwide have given high acclaim to the virtual authenticity of this enclosure. The graceful river otters are favorites at the museum and have inspired both poetry and prose. Lewis Thomas, renowned scientist, essayist, and National Book Award recipient, visited the museum and watched the swimming otters and beavers. In his book *Medusa and the Snail*, he described his reaction to the animals, "I was transfixed . . . there was only one sensation . . . pure elation mixed with amazement at such perfection." (Photograph by Mervin W. Larson.)

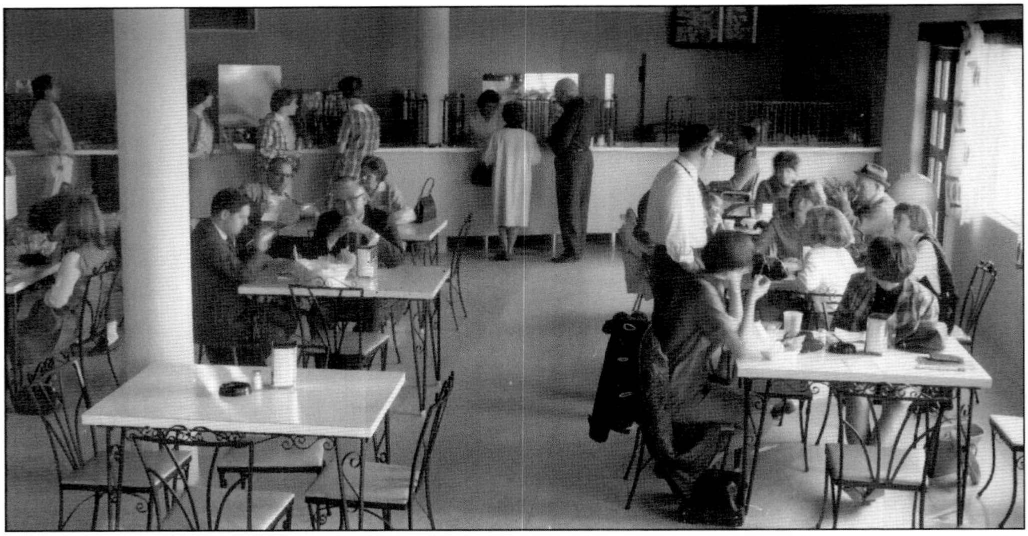

Service is in order in this refreshment area opened on Christmas Day 1965. Visitors are seen at the service counters patiently awaiting placed orders. The Mexican-style furnishings include the use of wrought iron tables and chairs graced with an application of turquoise. The museum established the dining facility shown here, and Mae Hogan, who with her husband had opened the first food concession at the museum, was in charge of the business. (Photograph by Mervin W. Larson.)

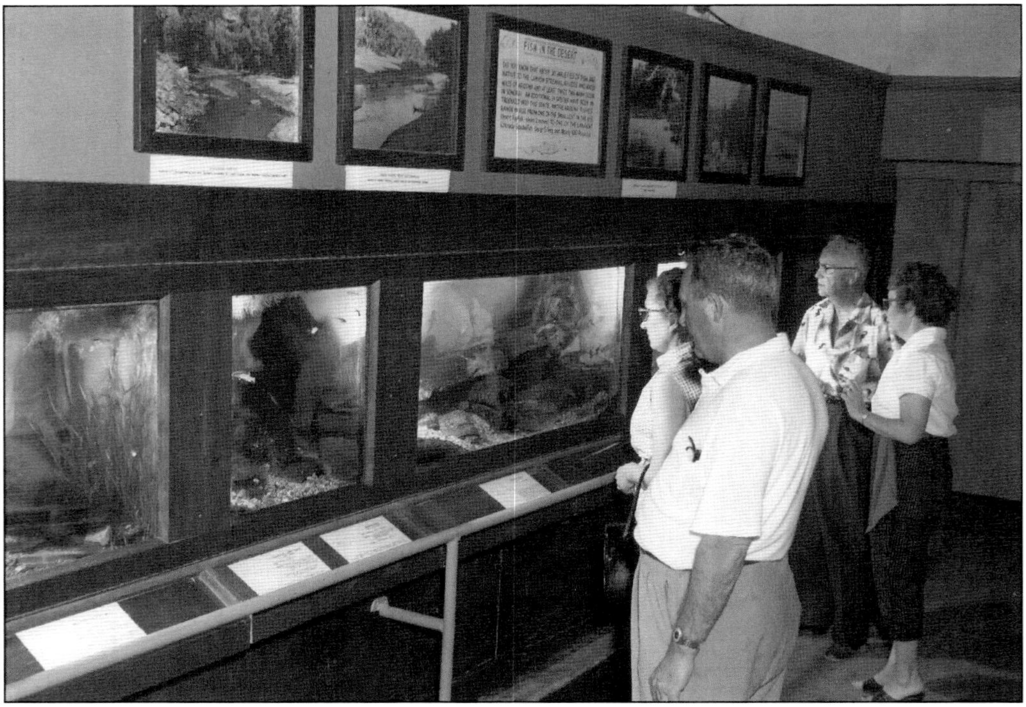

The Aquatic Room, shown here, includes indoor photographic labels lighted from the back. Many fish inhabit the riparian waterways of the Sonoran Desert. In this room, as early as the 1960s, the museum maintained one large salt-water tank that exhibited fish from the Gulf of California. (Photograph by Wyatt Davis.)

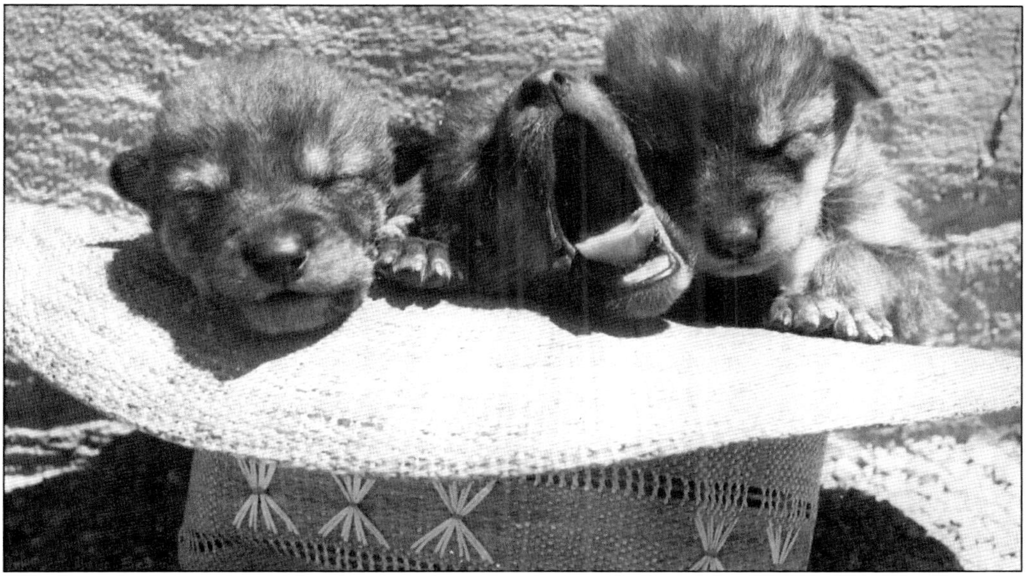

These three Mexican gray wolf pups were from a litter of six born at the museum in 1965. Two years earlier, the same female produced seven pups, which were the first Mexican gray wolves born in captivity. The mother of these two litters had originally come from Mexico; over her lifetime at the museum, she gave birth to a total of 46 pups. In the 1970s, the museum became a major partner involved with the United States Wildlife Service and the Mexican government in efforts to preserve and breed this species. (Photograph by Mervin W. Larson.)

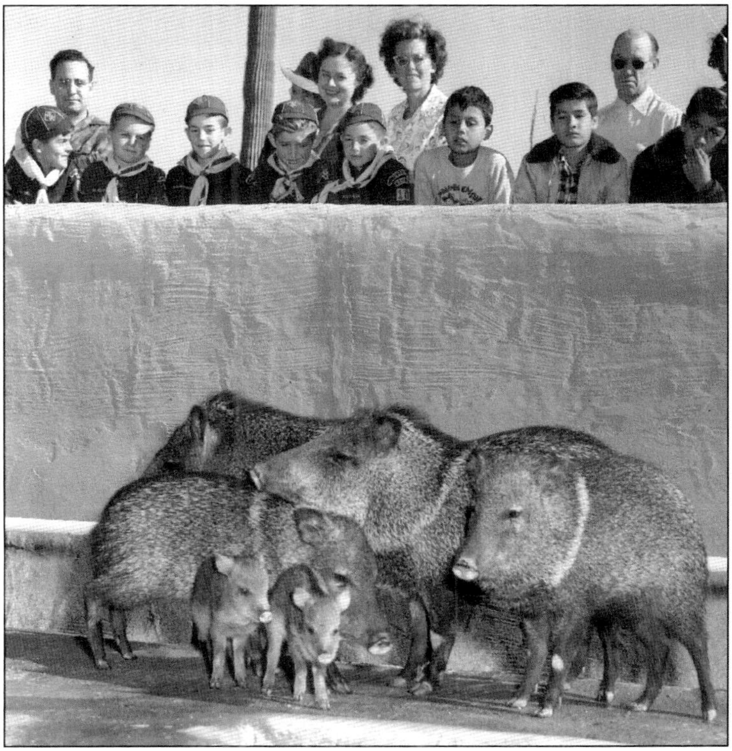

A small herd of javelina captivates an audience. Javelina live in small groups of 2 to 20 individuals and are common in the Sonoran Desert, where each herd defends a territory of about 700 to 800 acres. They consume plant material, including spine-covered prickly pear cactus pads. The adults have a coarse salt-and-pepper pelage, but the young are of a somewhat rusty-red color and are sometimes called "reds." Although they are often mistaken for pigs, they originated in the New World, and their anatomy also differs from that of pigs. (Photograph by Peter Balestrero.)

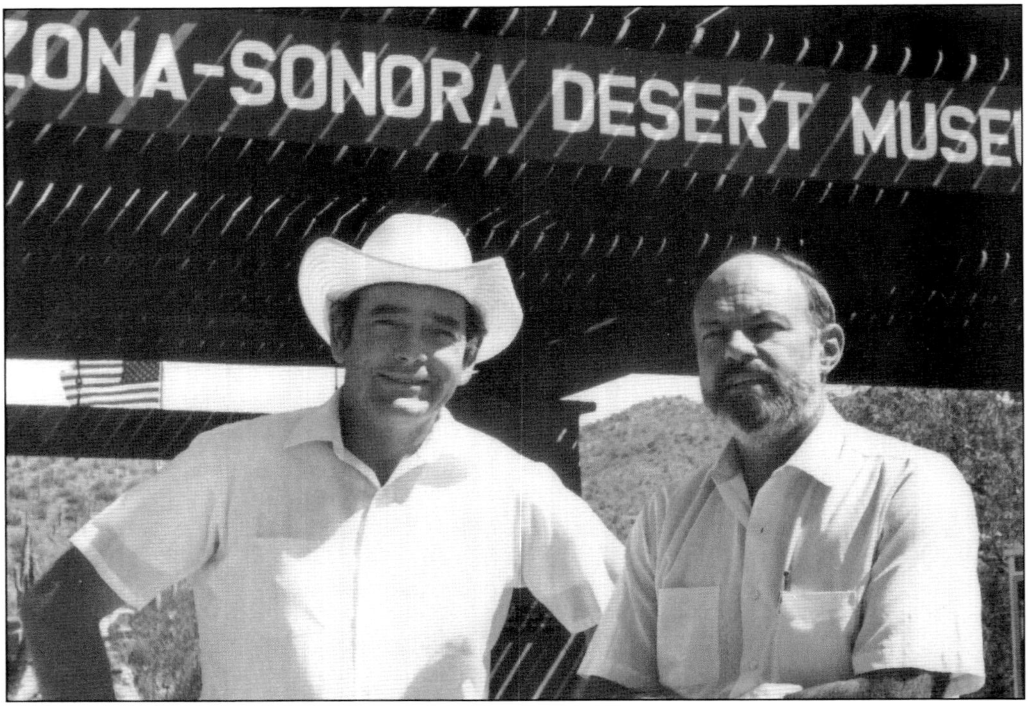

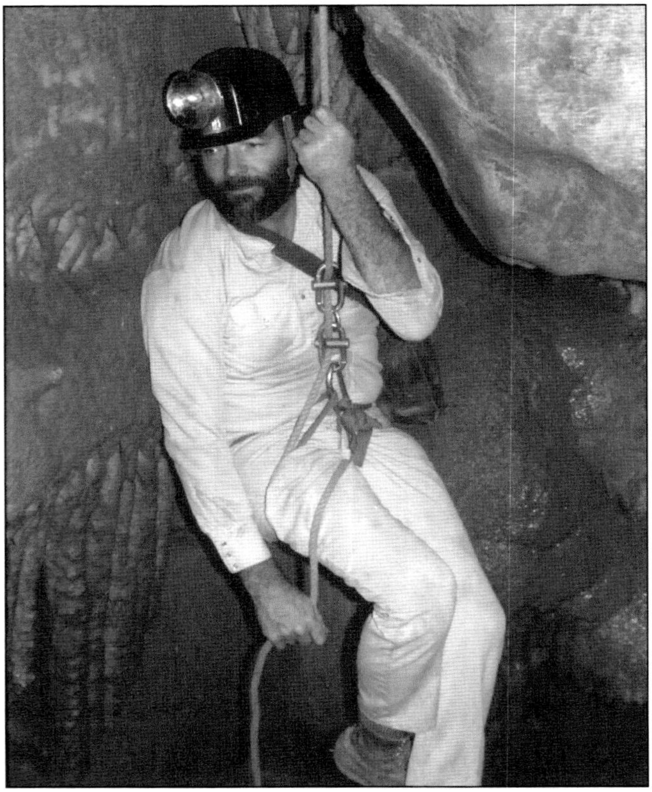

Bill Woodin (right) volunteered at the museum before its opening. He became director in 1954 and remained in that position for 17 years. Merv Larson began work at the museum in 1953 was later named curator of exhibits and general curator, then served as director from 1971 to 1976. Under their guidance, the museum became one of the preeminent institutions exhibiting live animals in the world.

In preparation for building the cave in the earth sciences center, staff members visited natural caves to photograph formations. Here director Merv Larson rappels into one on a trip organized by his son Lane Larson. (Photograph by Lane Larson.)

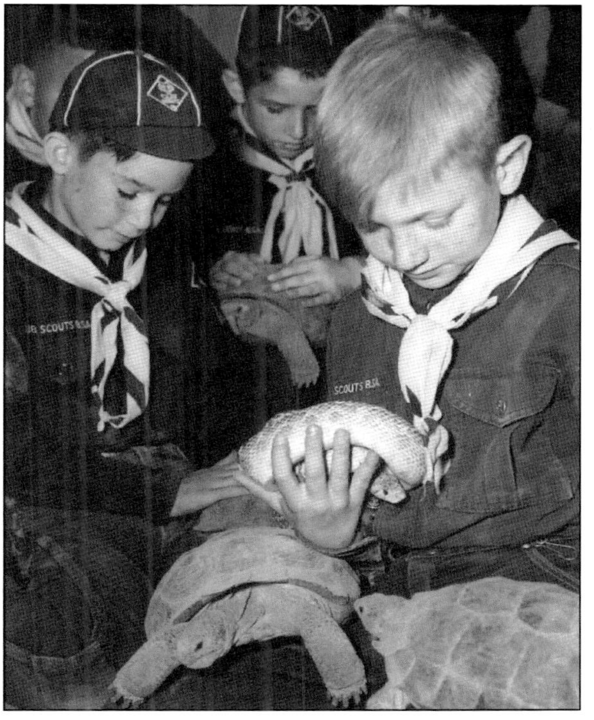

Hal Gras was doubtlessly present as Boy Scouts enjoyed the hands-on approach to Desert Ark wildlife at the museum. The desert tortoise and gopher snake appear not to mind the attention. (Photograph by Peter Balestrero.)

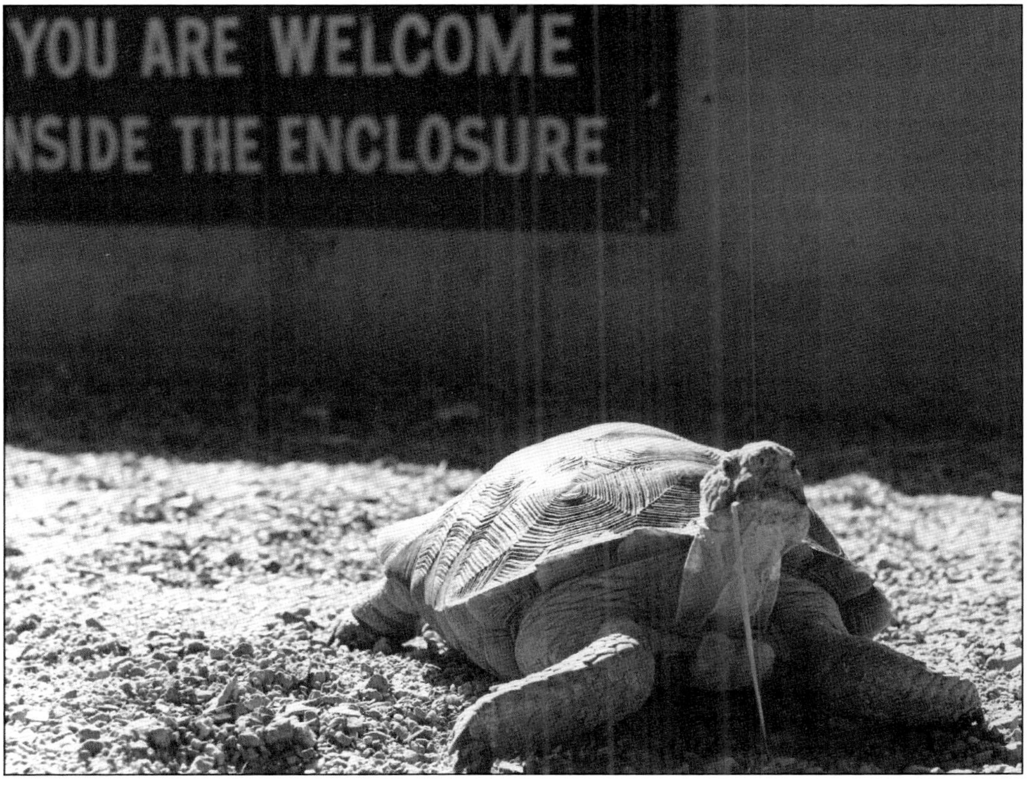

A low wall surrounded a large tortoise enclosure, and visitors, especially children, were urged to step over it and interact with a group of desert tortoises. The tortoises were not harmed, and today many adults fondly remember that experience at the museum.

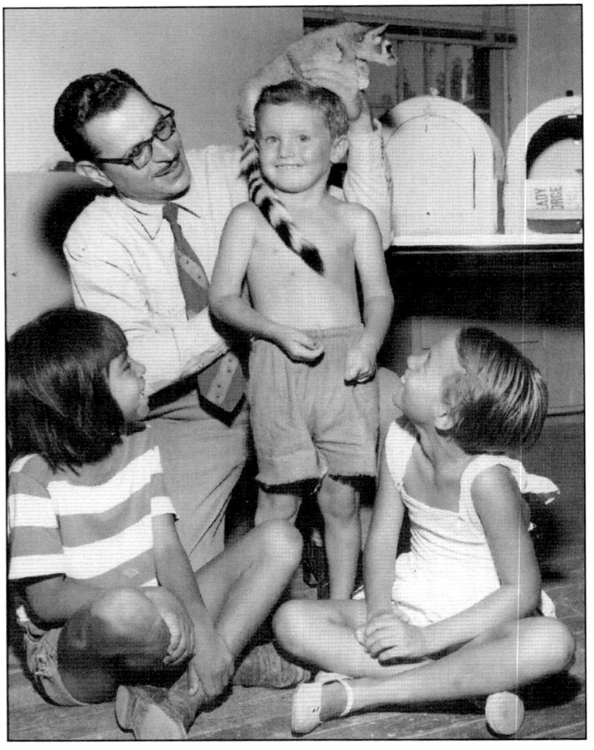

Hal Gras and La Vaga the ringtail shared many experiences with children at schools and on the museum patio on Sunday afternoons when Hal brought the Desert Ark animals to greet visitors to the museum. One of Gras's favorite acts was to place La Vaga on a child's head, giving them the impression of wearing Davy Crockett's hat. The kids loved it, and in her 14 years filling that role, La Vaga never complained.

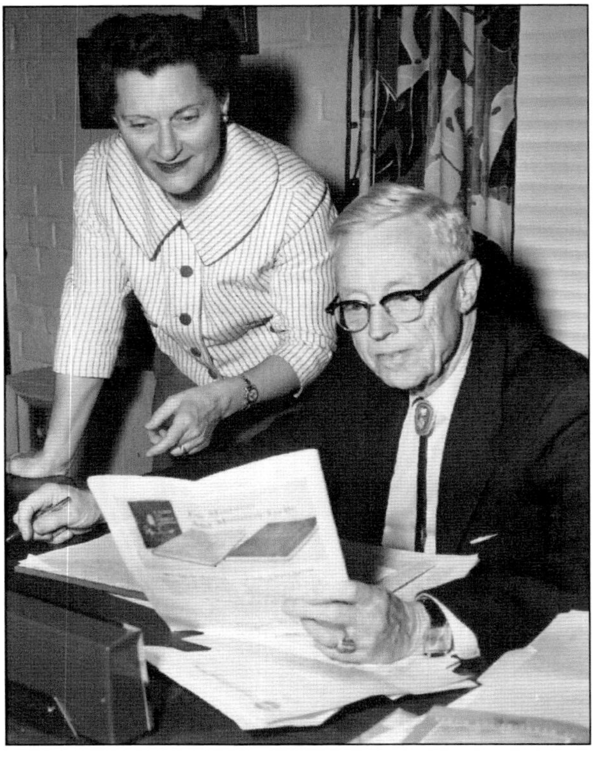

Phoebe and Arthur Pack were philanthropists and active community members in Tucson. Arthur Pack had established Ghost Ranch in northern New Mexico. When he and Phoebe moved to Tucson in the late 1940s, they established Ghost Ranch Lodge. Phoebe Pack was particularly active in volunteer work in Tucson hospitals, and Arthur participated in and helped subsidize many community projects, including the Desert Museum. The couple founded Ghost Ranch Museum, similar to the Desert Museum, near their original ranch and donated an entire hospital for use by the rural inhabitants living near Ghost Ranch in New Mexico.

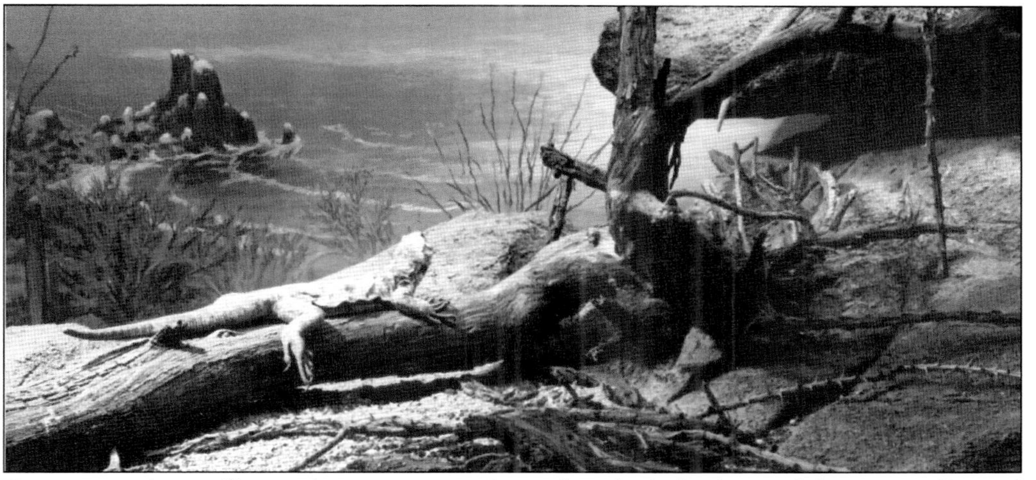

Dioramas in the small animal room were similar to those built for the amphibian room. Painted on the background pictured here is the Sea of Cortez with San Esteban chuckwalla lizards in the foreground. In the Sea of Cortez, a few islands harbor gigantic chuckwalla lizards that are much larger than those on the mainland. The San Esteban chuckwalla occurs only on the one island, and since 1977 the Desert Museum has been active in studying and breeding these lizards to serve as a genetic reservoir for the species. The Seri Indians formerly used chuckwallas as emergency food when stranded on islands in storms, but now they work to protect these animals. In the late 1990s, the Desert Museum staff, in conjunction with the Seri tribe, built a chuckwalla breeding enclosure in Punta Chueca, Sonora, where museum personnel trained Seri residents in chuckwalla management.

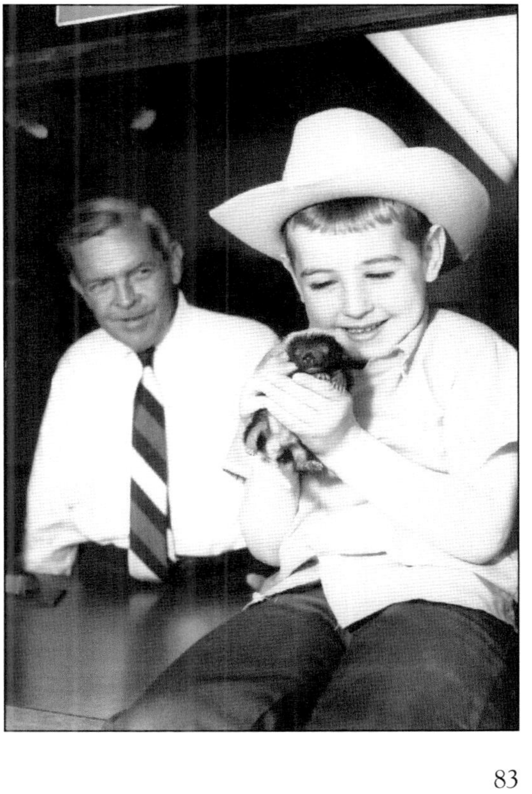

Five-year-old David Moote from Kalamazoo, Michigan, is being rewarded by the Arizona-Sonora Desert Museum for being its three-millionth visitor. Not only did he receive a lifetime membership to the museum, he also made a friend, according to this photograph taken May 22, 1968. The baby hognose skunk he is holding is only a month old and appears to be unconcerned about the attention surrounding him. William O. Morgan, admissions and assistant treasurer, looks on.

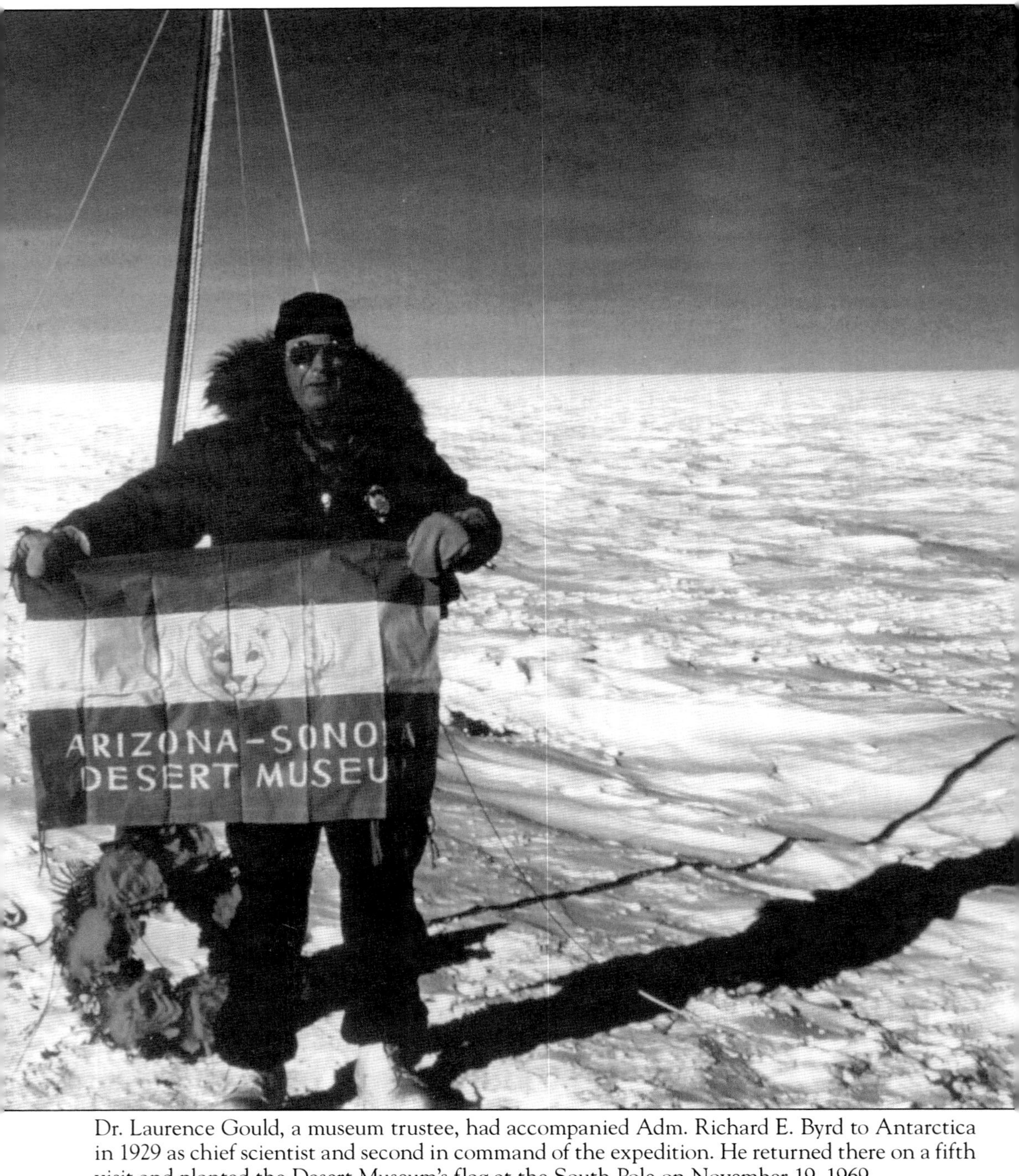

Dr. Laurence Gould, a museum trustee, had accompanied Adm. Richard E. Byrd to Antarctica in 1929 as chief scientist and second in command of the expedition. He returned there on a fifth visit and planted the Desert Museum's flag at the South Pole on November 19, 1969.

Three

1970s

In the early part of the 1970s, the museum flourished with creativity and implementation of new techniques in exhibitory and increased emphasis on education, volunteerism, conservation, and extensive outreach to schools and institutions in Mexico.

Exhibits started in the late 1960s came to fruition in the 1970s. The final phase of the Demonstration Desert Garden opened in 1971. The beaver-otter-sheep complex was also finished in 1971. The habitat for small cats opened in 1973. The Stephen H. Congdon Earth Sciences Center was started mid-decade and the wet and dry caves completed during that time.

Lewis Wayne Walker retired January 1, 1970; Bill Woodin retired two years later. In 1971, Merv Larson was named director designate and became director upon Woodin's retirement. The Education Department was formed with Doris Ready as head, under which an outstanding docent program was implemented. Increased numbers of classes and study trips were organized, including one for 13 teenagers on a three-week study and camping trip in Baja, California, and another desert-study trip to South Africa and Namibia. Nick Wilson, staff artist, painted outstanding museum posters depicting desert wildlife.

A new walk-through aviary was built. The old Large Animal Line was demolished in preparation for building the Mountain Habitat. A successful breeding program for endangered margays and ocelots was initiated. Tommy, the museum's first bobcat, died at the age of 21, and El Tigre, the first jaguar, also died at age 21. The museum's female Mexican gray wolf died after she had given birth to a total of 46 pups during her long museum stay. Chivito, the museum's first bighorn sheep, arrived as a lamb and charmed visitors. The beautiful mountain lions Honey and Tom came to the museum, and the death of George III after 17 years resulted in Tom becoming George IV. The mountain lions Honey and Georgette gave birth within two days of each other to a total of six cubs. These were the first born at the museum, and a local newspaper proclaimed, "Old George L. would be proud."

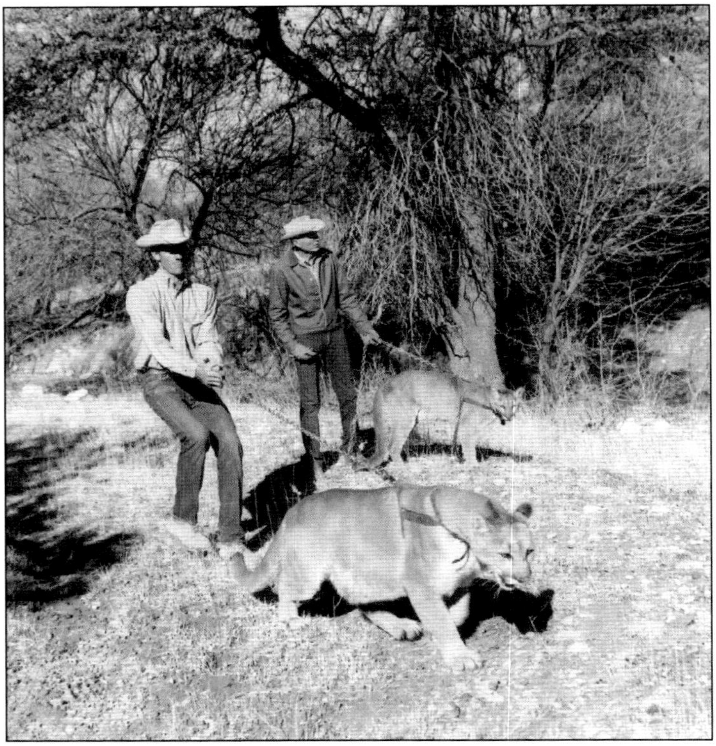

Tom (foreground) and Honey, the mountain lions, are shown here being walked by Warner (left) and Marvin Glenn on their family ranch in southeastern Arizona. The lion brother and sister, born and orphaned in 1969, were raised on the ranch until they were donated to the museum in February 1971. Honey, particularly, was tame and often taken for walks on the museum grounds. Tom eventually became George L. Mountainlion IV. (Photograph by Mervin W. Larson.)

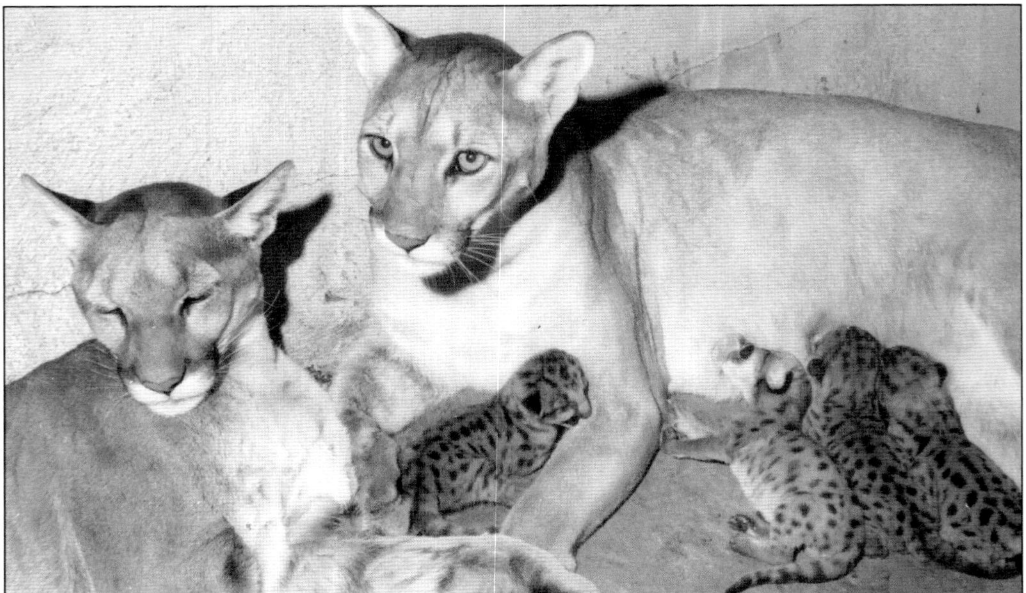

Honey gave birth to two kittens, and Georgette produced four just two days later. The two mothers were in the same enclosure, and the kittens nursed from both with no objections from the two females. The adult lions allowed museum personnel with whom they were familiar to enter their quarters and handle the young. (Photograph by Mervin W. Larson.)

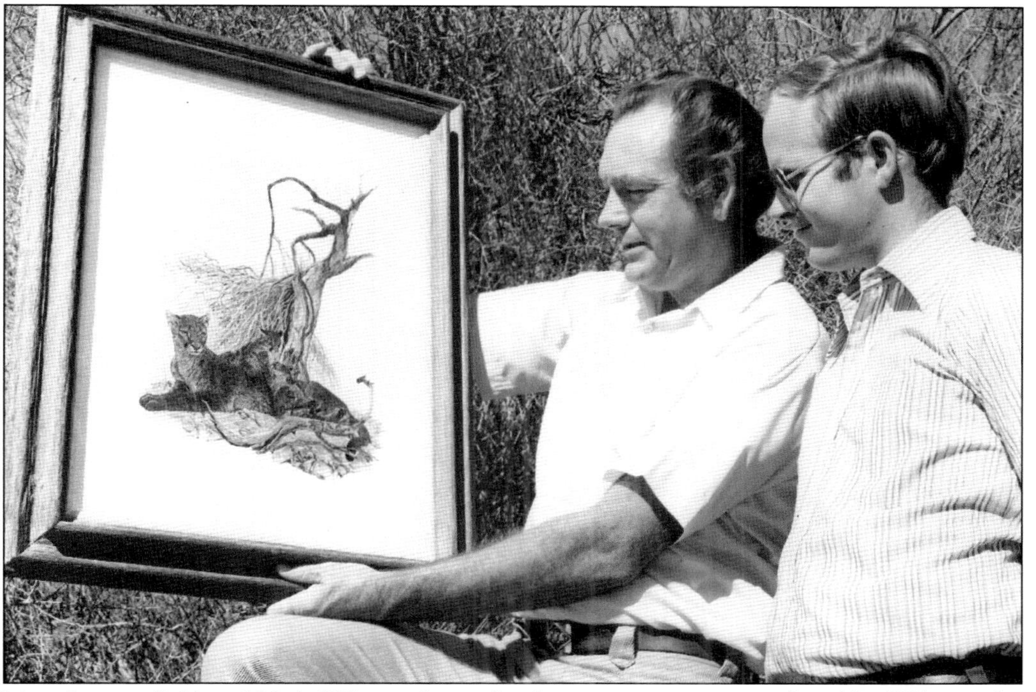

Merv Larson (left) and Nick Wilson admire the first museum poster produced by Wilson, who served as staff artist for the museum. His renowned artwork appeared on several early museum posters, and he was responsible for designing the museum's lion logo.

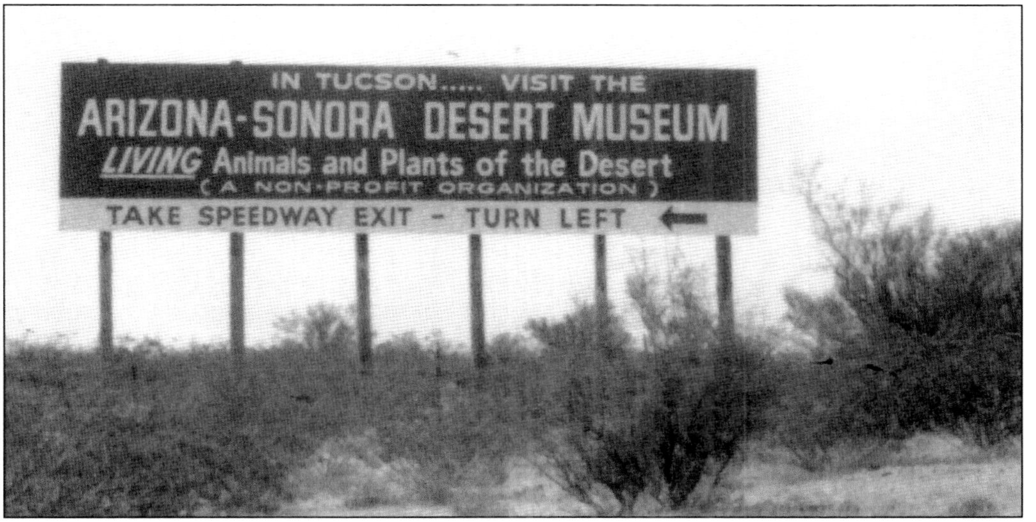

This billboard depicts a March 1972 Arizona-Sonora Desert Museum advertisement off Interstate 10, some 10 miles east of Tucson on the left-hand side of the road approaching Tucson. There was considerable debate among board members regarding the advisability of the museum advertising through billboards in the 1960s. Ultimately several were used on the approaches to Tucson. Their value, or lack of value, was never clearly determined. With popular opinion gradually becoming negative toward the presence of billboards along highways, their use by the museum was terminated. (Photograph by Richard F. Sowers.)

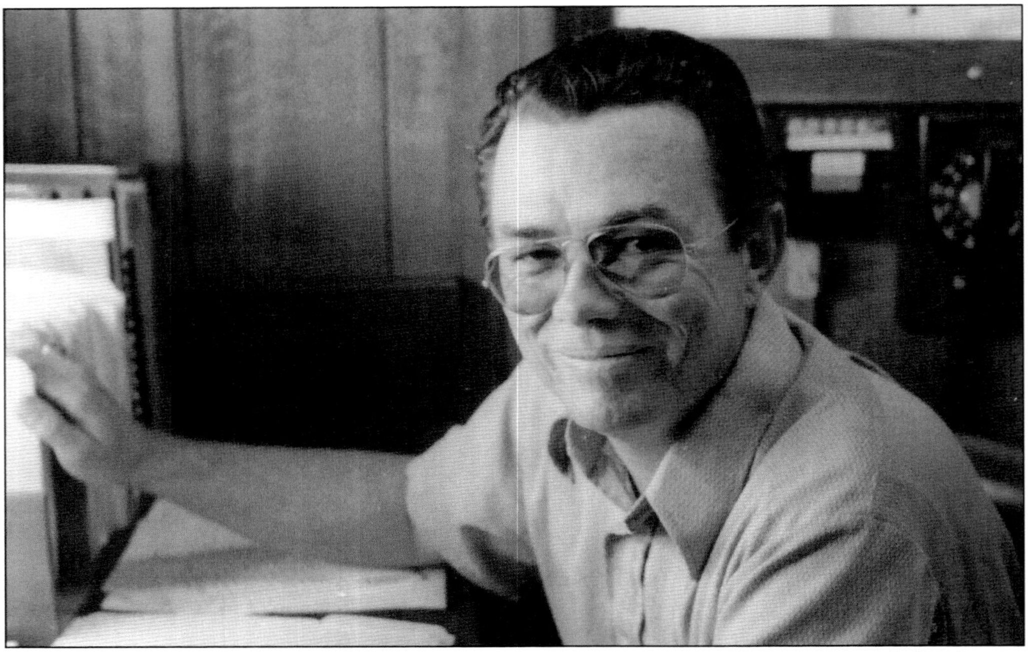

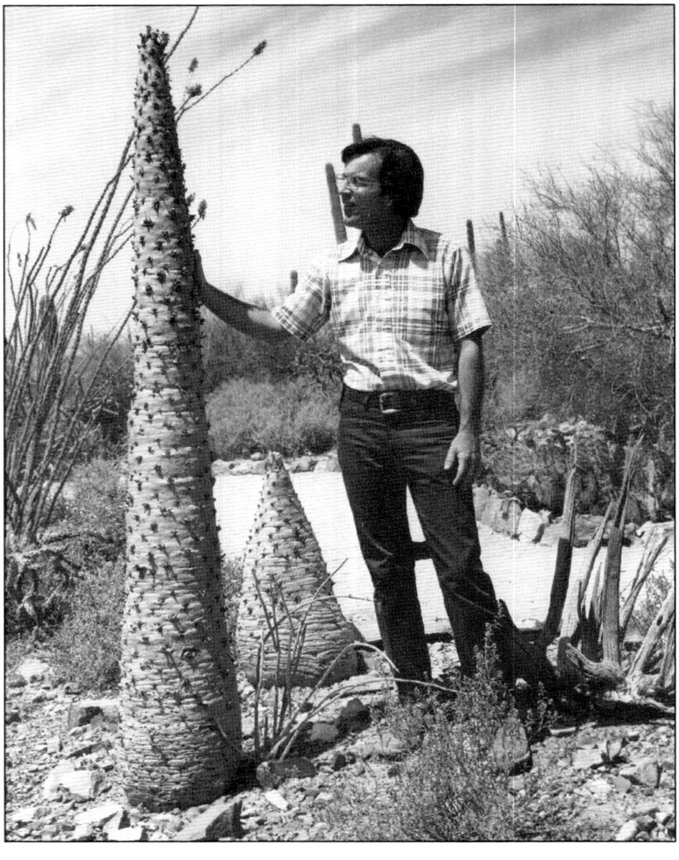

Robert M. Craig helped in construction of the Tunnel in the 1950s. Later, as an employee of the University of California at Davis, he was of assistance in the museum's bobwhite quail conservation project. He returned to the museum and served as general curator in the 1970s.

Curator of plants Don Ducote proudly displays a boojum tree on the grounds of the Arizona-Sonora Desert Museum. This plant is a near endemic to central Baja, California, with a small disjunct stand on the mainland coast of Sonora. The plant is an eye-catcher for the public, who characterize it as an upside-down carrot. Obtaining a height of over 60 feet, this spiny-stemmed relative of the more familiar ocotillo produces small white fragrant flowers in the late summer.

An animal nursery was established in the 1970s at the museum and proved very popular with visitors, who could observe nursery activities through a large window. Young animals, including bobcats, foxes, and occasional mountain lions, were numerous in the spring and early summer, and their care in the nursery was largely provided by docents supervised by staff. Here Dagmar Sommer, a large animal keeper, is shown in the nursery with a young kit fox and provides a good example of the close-knit, keeper-animal relationships common at the museum.

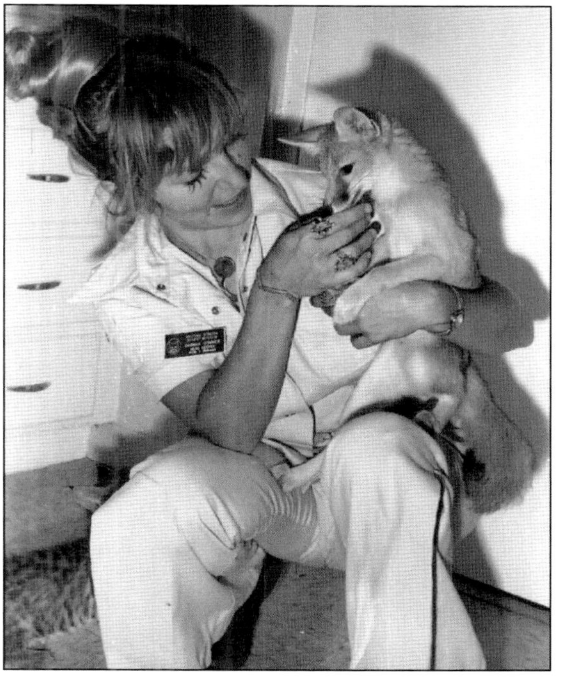

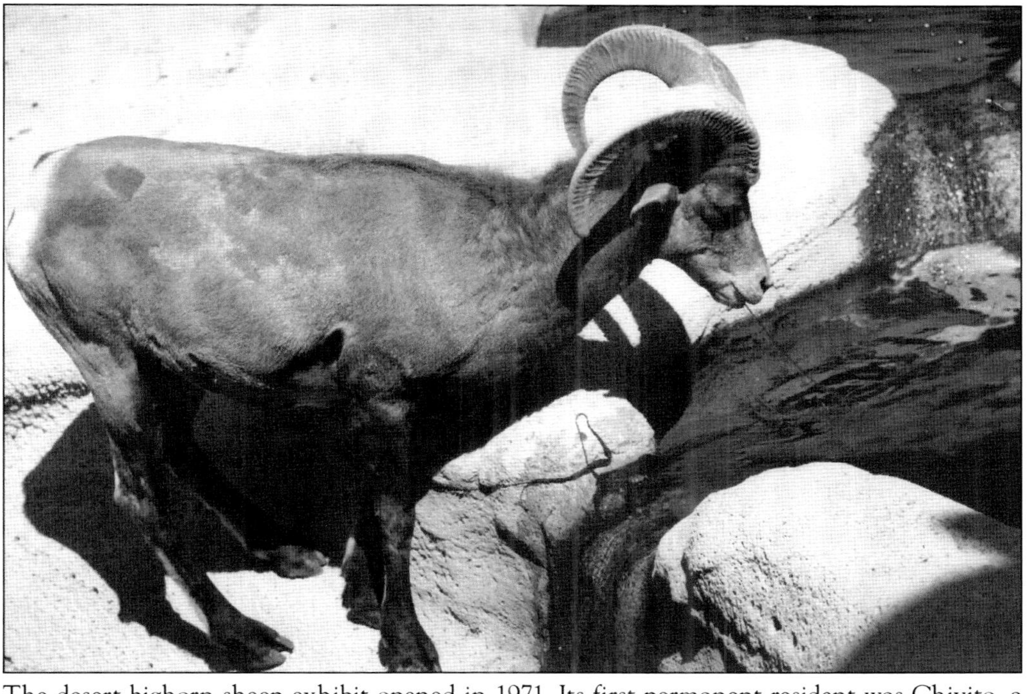

The desert bighorn sheep exhibit opened in 1971. Its first permanent resident was Chivito, a young lamb born in Mexico and donated to the museum by a rancher who rescued the orphan. The bighorn enclosure is modeled after the White Tanks area in central Arizona, and steep walls of naturalistic rock on a variety of levels provide the sheep with a habitat similar to areas where they live in the wild.

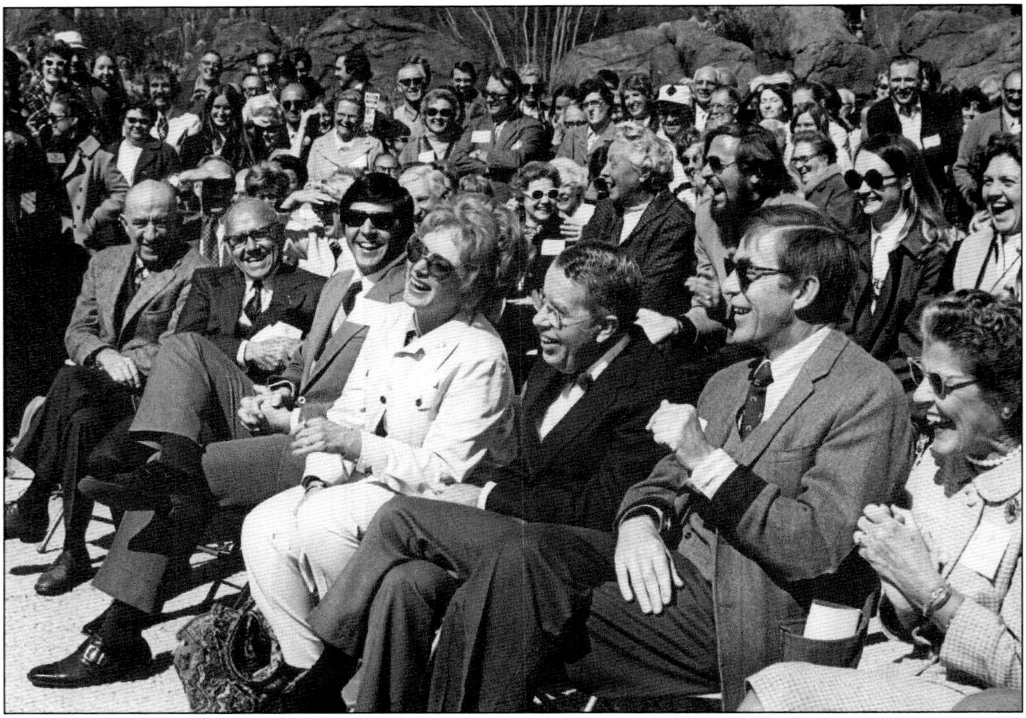

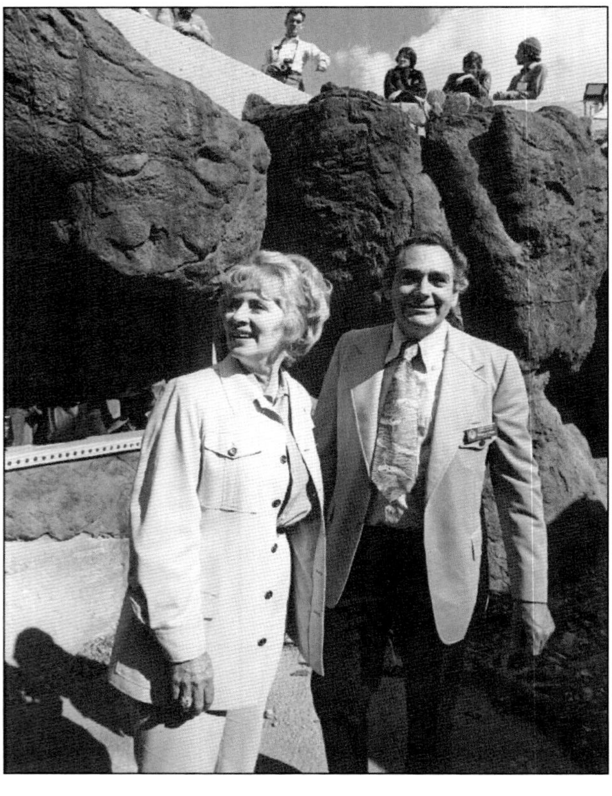

The Mary Magee Scaife Memorial Canyon Habitat for Small Cats, dedicated on March 10, 1973, has hosted a number of species throughout the years, primarily bobcats, margays, ocelots, and jaguarundis. The rock canyons are naturally sculpted with concrete and molded to steel frames. Amanda Blake, the renowned actress who portrayed "Miss Kitty" on the popular western television show *Gunsmoke*, served as a member of the board of trustees. She is flanked by Arizona governor Jack Williams on her left at the dedication ceremony seen above. A strong supporter of the Arizona-Sonora Desert Museum and active in the conservation of wild cats worldwide, she is seen below with Director Larson as a special guest at the dedication of Cat Canyon.

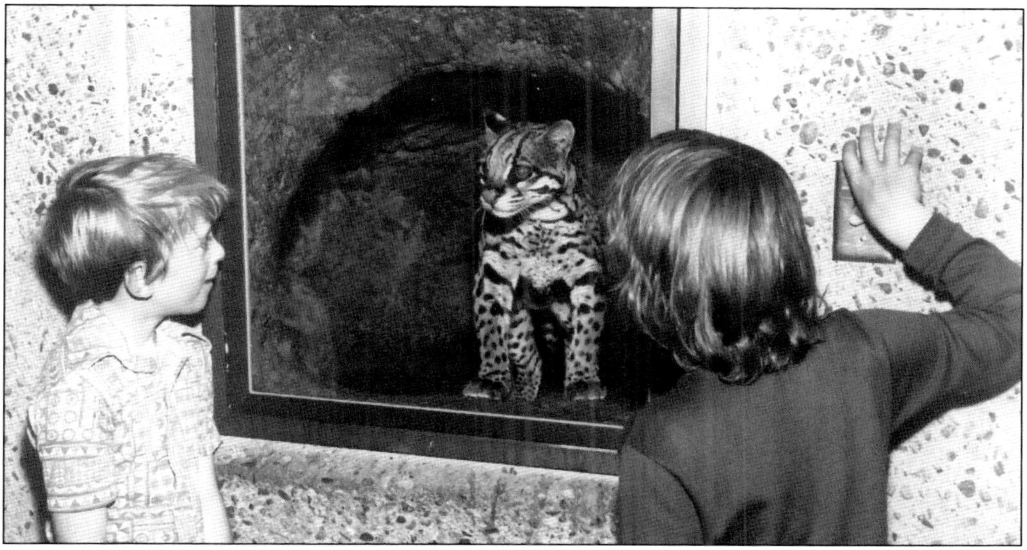

Children are enjoying an up-close view of an ocelot in Cat Canyon. Viewing windows such as these provide visitors a personal experience with these rare endangered cats. Visitors can view the cats from several viewpoints on a variety of levels in the Cat Canyon exhibits. The retinas of many nocturnal animals contain a tapetum lucidum layer, which reflects light, enhancing night vision.

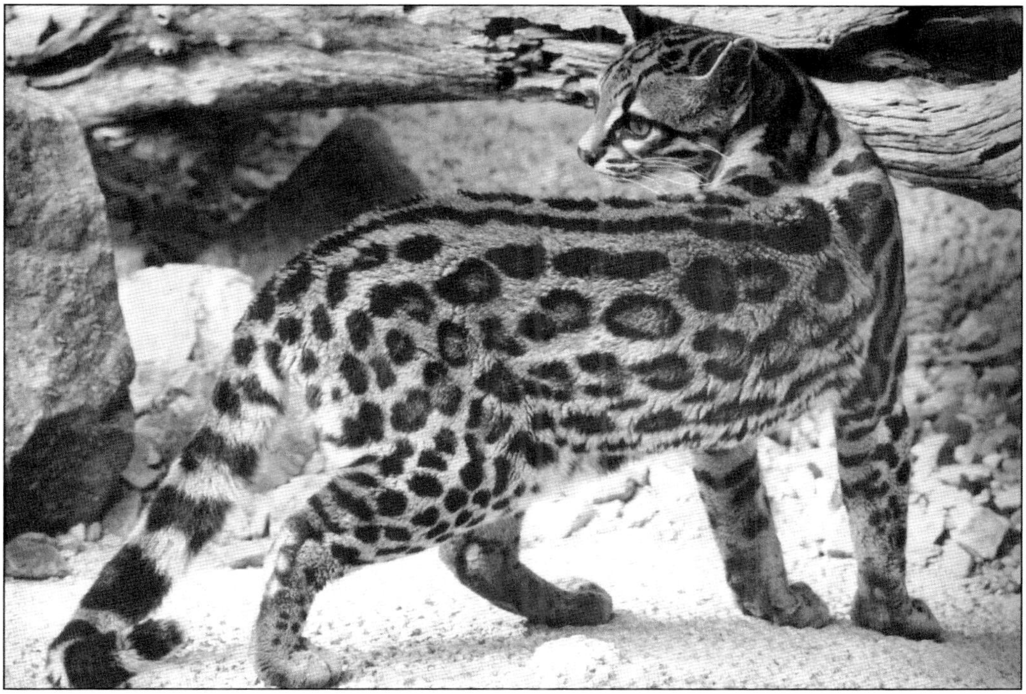

Renowned for its beautiful coat composed of dots and stripes, the ocelot may still very occasionally occur along the border in southeastern Arizona, one having been recorded there by photograph in 2010. In the 1970s, the museum conducted a highly successful breeding program for both ocelots and margays. (Photograph by Verna R. Johnston.)

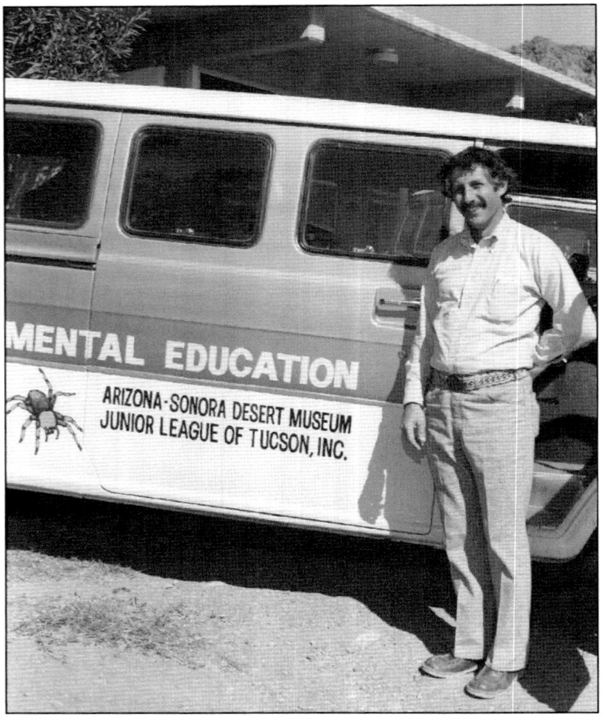

Carlos Nagel coordinated the Desert Museum Mexico Unit created in 1974. This program established education and conservation outreach between the two bordering countries. The Desert Museum served as a consultant when the Centro Ecológico de Sonora was being planned and established in Hermosillo, and cooperation between the two institutions in key research projects has been ongoing.

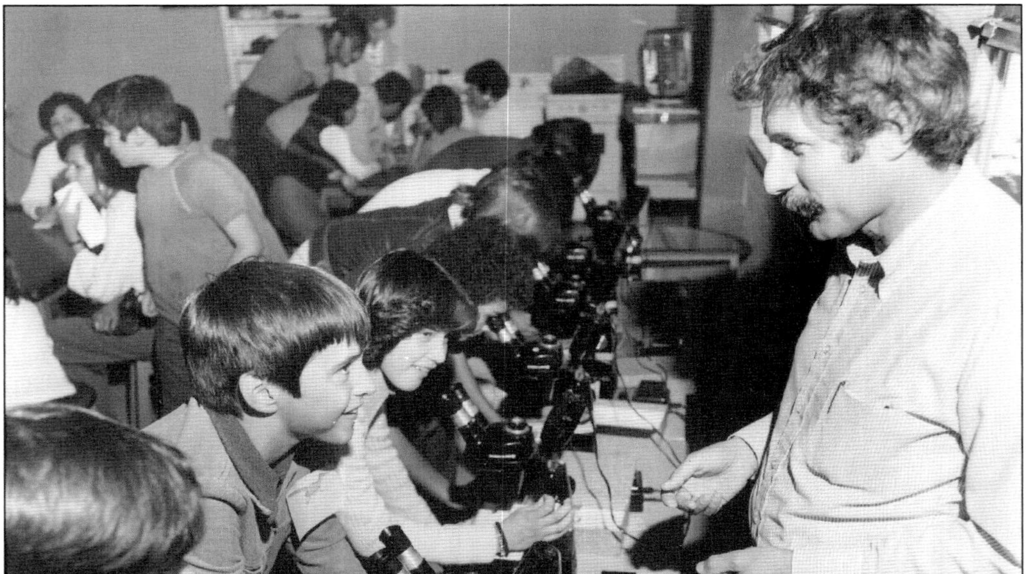

Carlos Nagel is seen here working with children in Sonora, Mexico. Nagel, Doris Ready (the museum's director of education), and Merritt Keasey (the curator of small animals), traveled to Sonora working with teachers and students in science education. The museum was awarded the American Association of Zoological Parks and Aquariums' Special Achievement Award, which recognized years of work in bilingual environmental education with the schoolchildren of Sonora, Mexico.

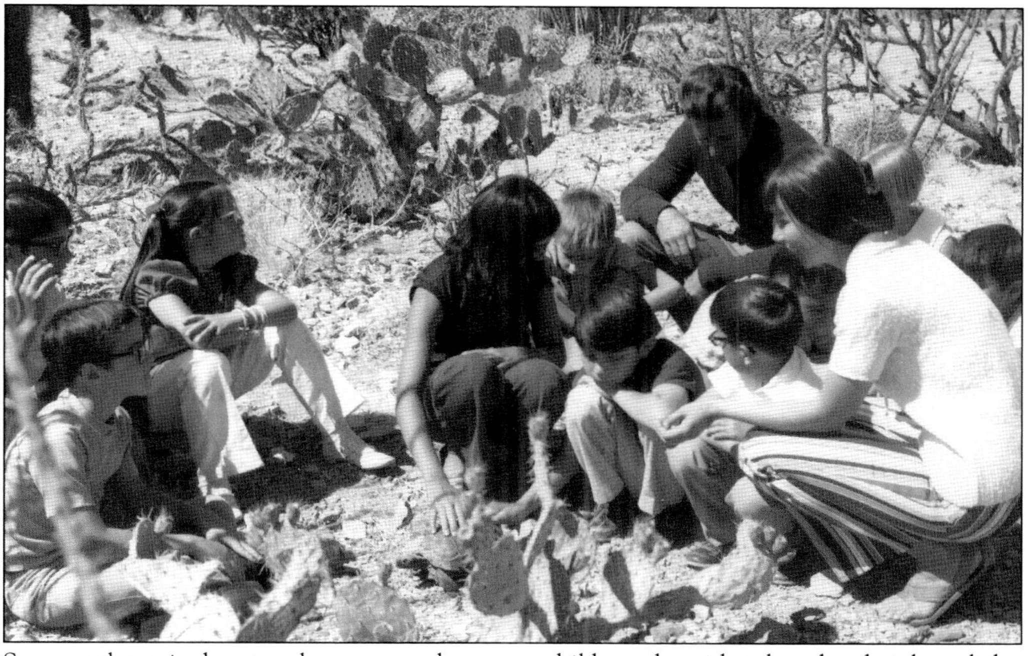

Summer classes in desert ecology are popular among children who wish to broaden their knowledge of the Sonoran Desert. Doris Ready (far right), director of education, is shown here enlightening an audience of children on the vegetation surrounding the Desert Museum. (Photograph by Fran Coffey.)

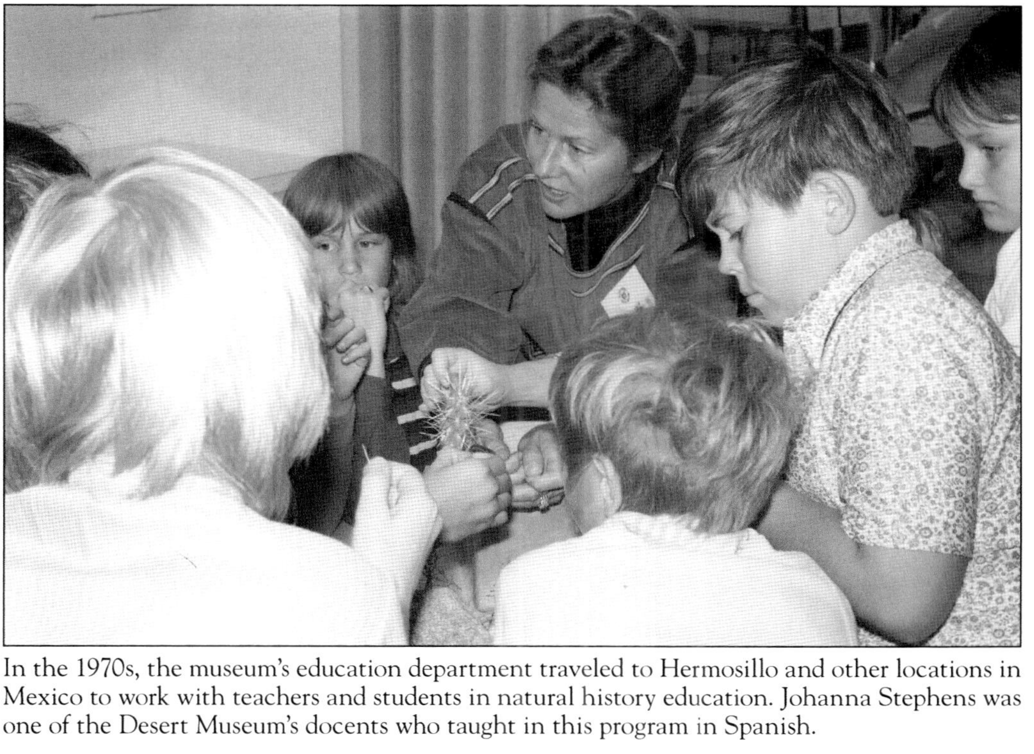

In the 1970s, the museum's education department traveled to Hermosillo and other locations in Mexico to work with teachers and students in natural history education. Johanna Stephens was one of the Desert Museum's docents who taught in this program in Spanish.

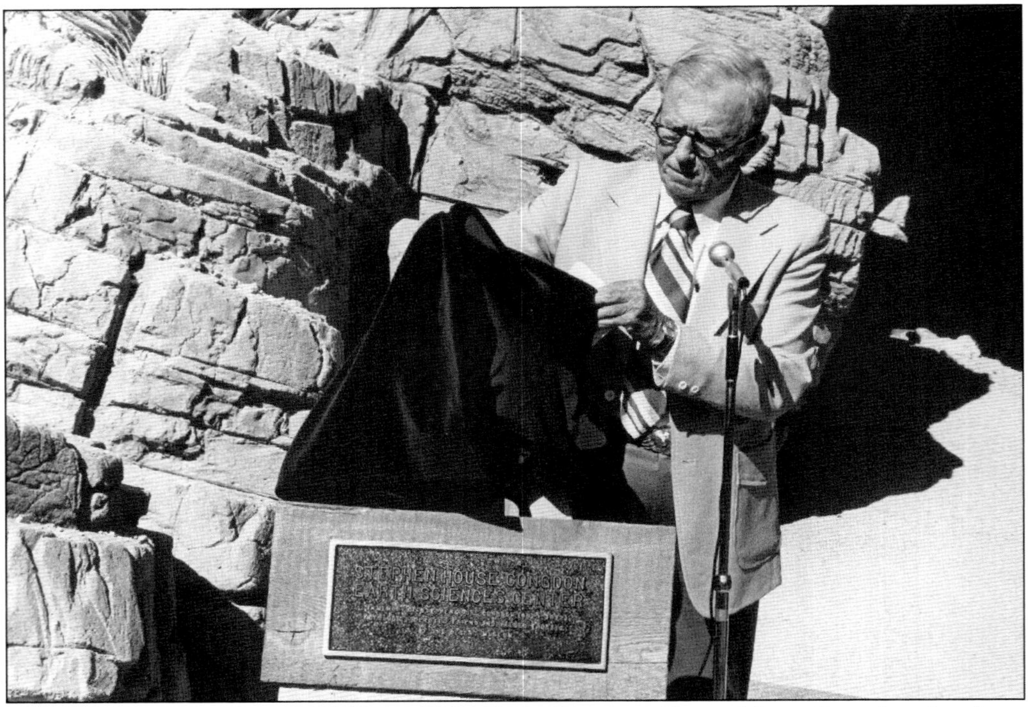

Here trustee Dr. Laurence Gould unveils the plaque honoring Steven House Congdon, for whom the Earth Sciences Center is named. Stephen House Congdon was a geologist and valued museum trustee who died in 1973. His family was instrumental in establishing the Earth Sciences Center and dedicating it in his name.

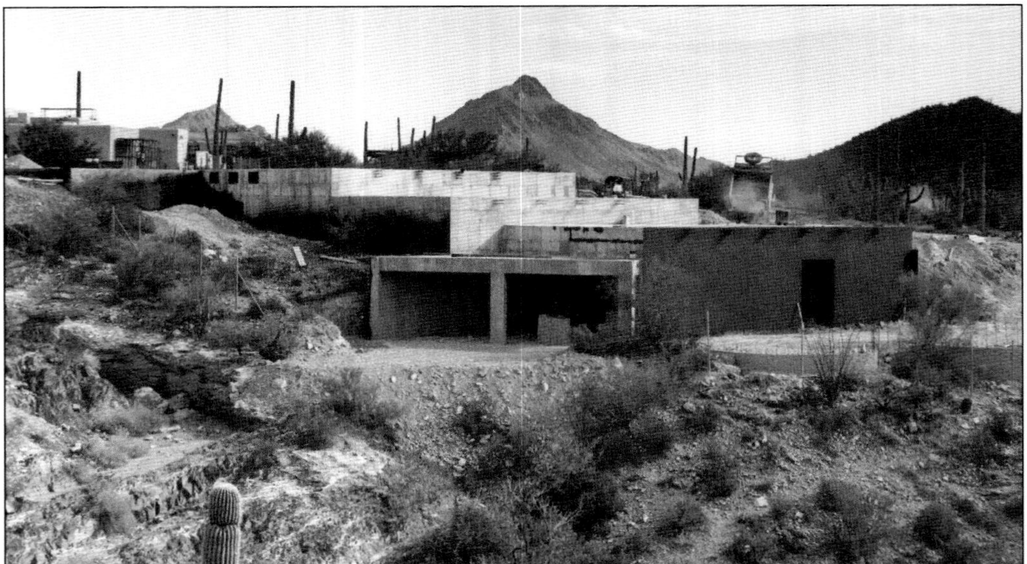

Construction is well underway in the creation of the Stephen H. Congdon Earth Sciences Center. Note how Golden Gate Mountain stands majestically in the middle background with Brown Mountain to the right.

The manmade wet and dry caves at the Stephen H. Congdon Earth Sciences Center are notable for their realistic stalactites and stalagmites carved by water in limestone. The caves were recognized for their authenticity by the American Concrete Institute.

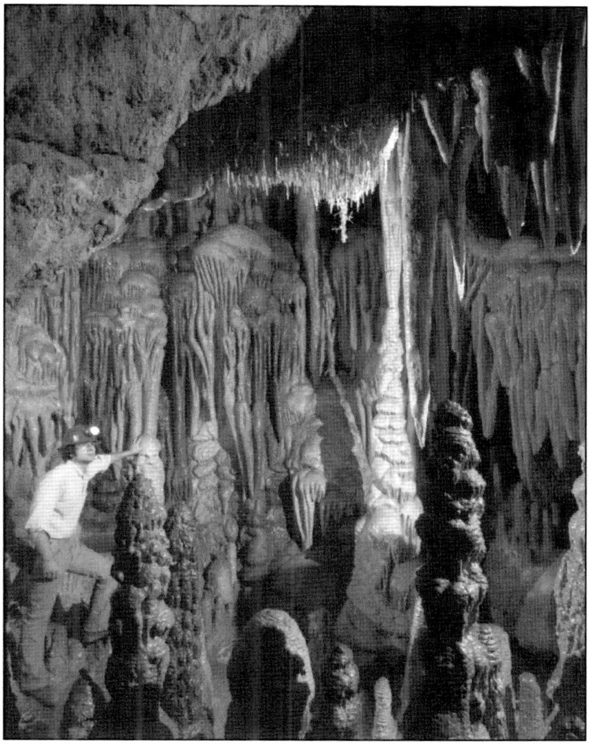

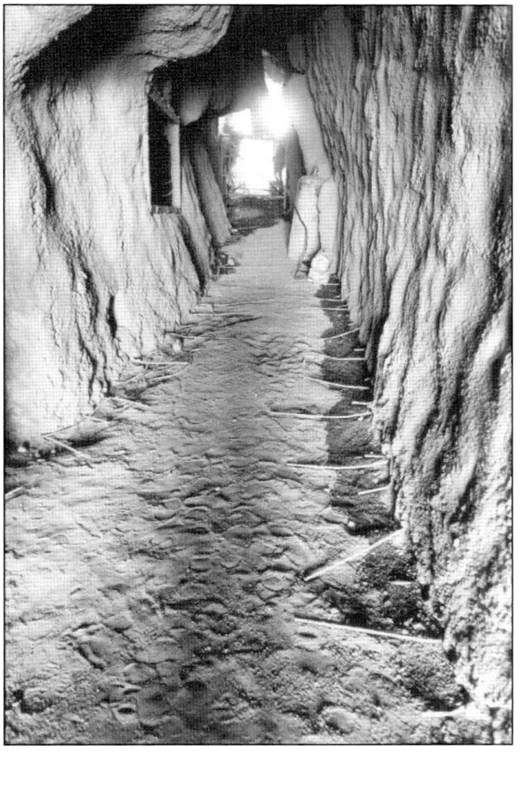

Here under construction, the 67-foot tunnel in the Stephen H. Congdon Earth Sciences Center was finished to represent Earth's timeline from nearly five billion years ago to the present. The presence of man represents less than three-eighths of an inch of the 67 feet.

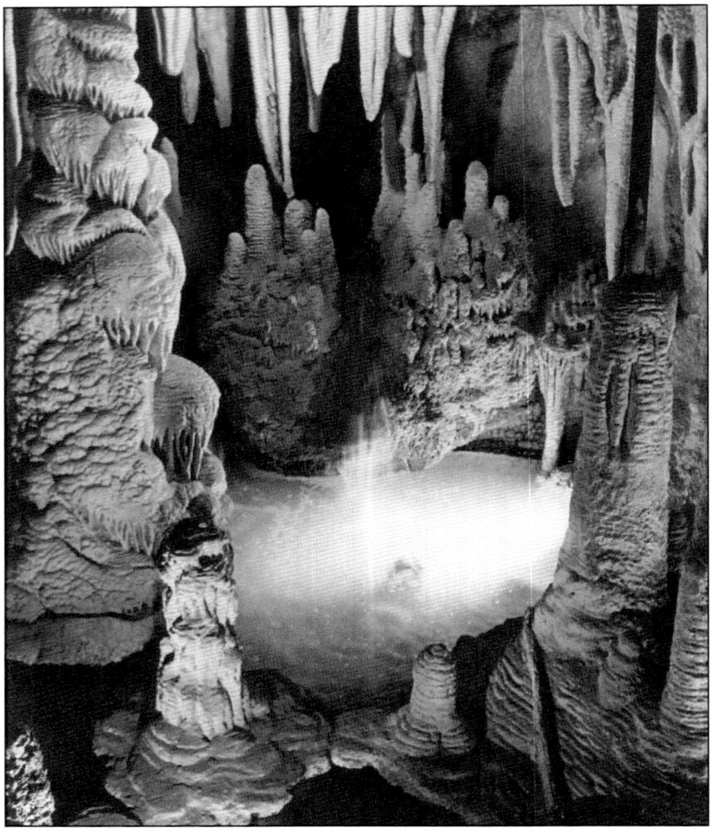

The cave pool helps complete the wet underworld of the manmade cave in the Stephen H. Congdon Earth Sciences Center. Limestone and water reacting over millions of years created many of the wet caves worldwide. (Photograph by Al Morgan.)

Flowstone and columns are specially crafted features that decorate the inside of the Stephen H. Congdon Earth Sciences Center. A narrow passageway allows those who wish to venture through the recesses of this manmade limestone cave an opportunity to explore. (Photograph by Al Morgan.)

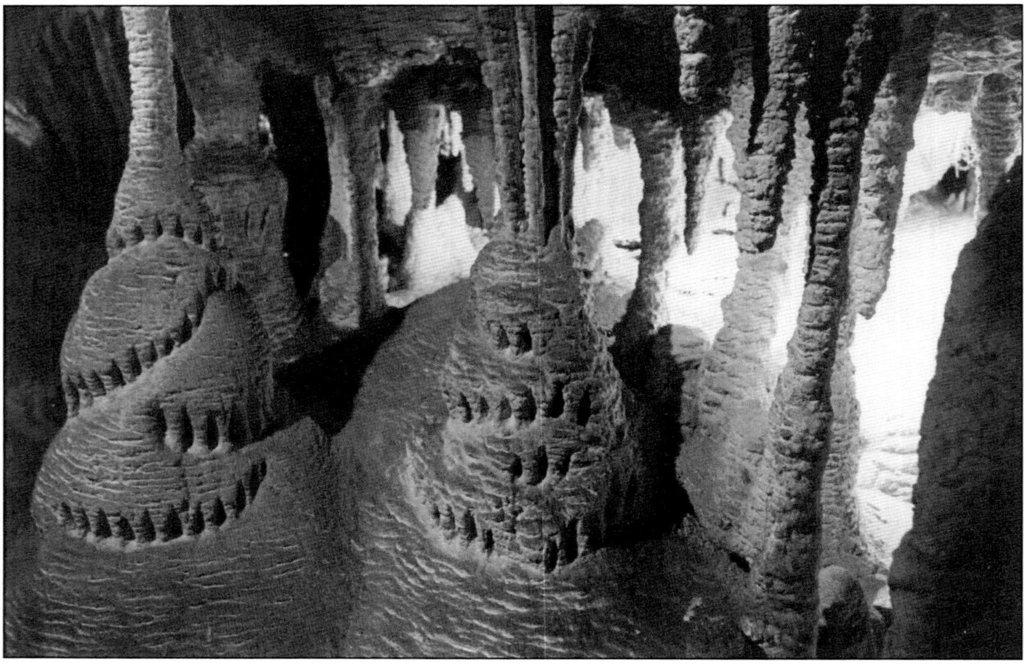

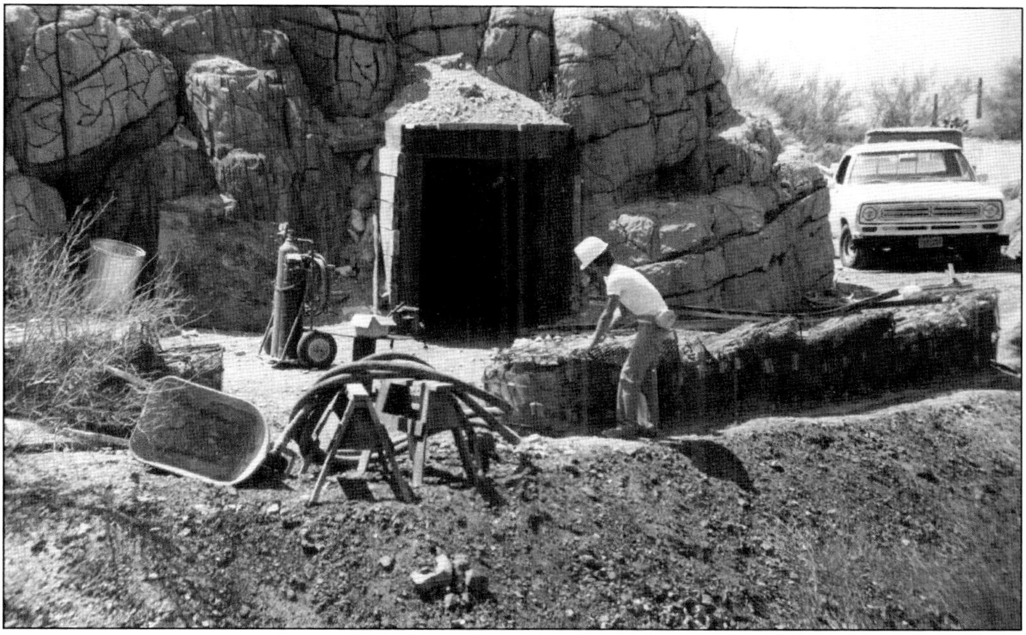

Construction at the mine entrance of the earth sciences center is depicted in this photograph. Some of the lagging or timbering was acquired from Asarco's Groundhog Mine at Vanadium, New Mexico.

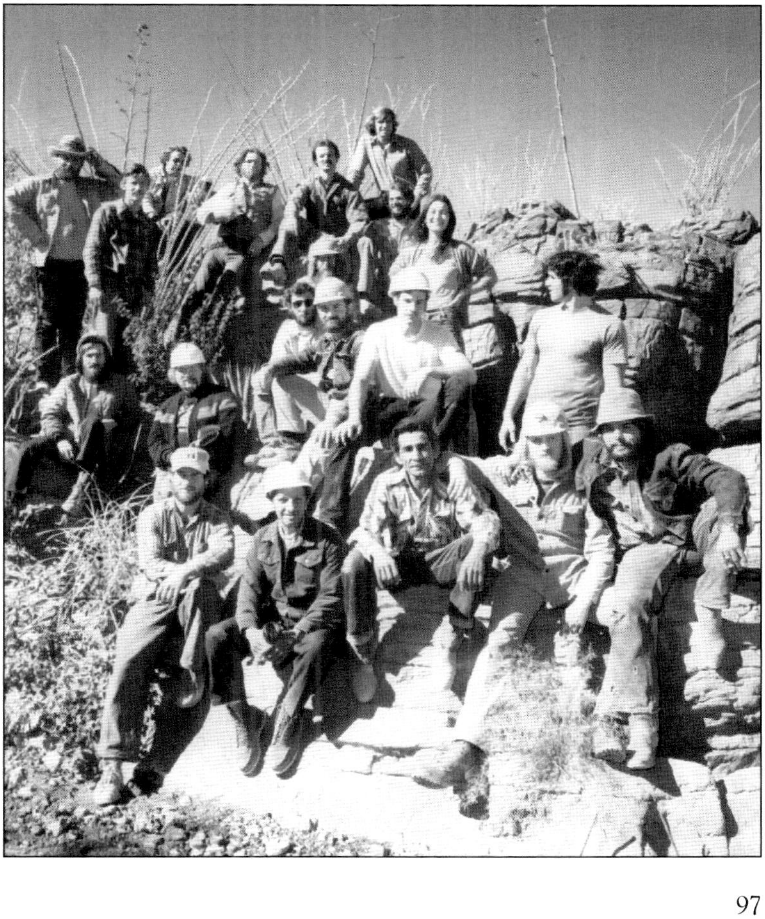

Building naturalistic exhibits is obviously labor intensive. The early construction crew of the cave was a large one, as is evident in this photograph.

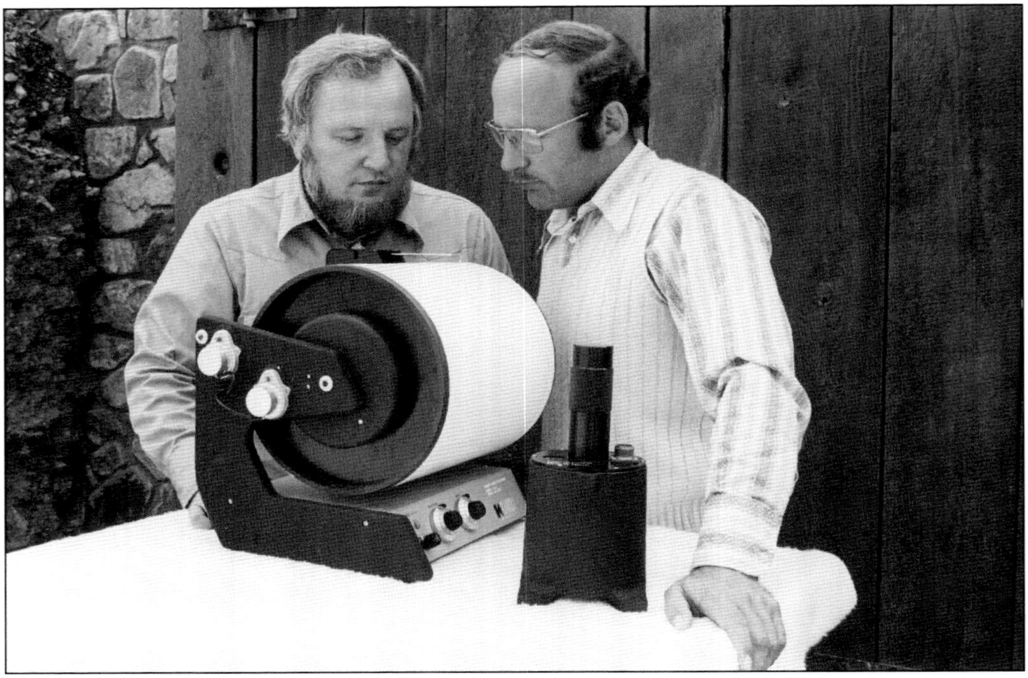

Bill Panczner (left) and Peter Kresan, former president of the board of trustees, are discussing the merits of the new seismograph in Earth Sciences. Bill Panczner became the first curator of Earth Sciences in 1974. During the course of the next decade, the museum's earth sciences department would be at the forefront of geologic interpretation with pioneering exhibits such as the artificial cave and mineral room.

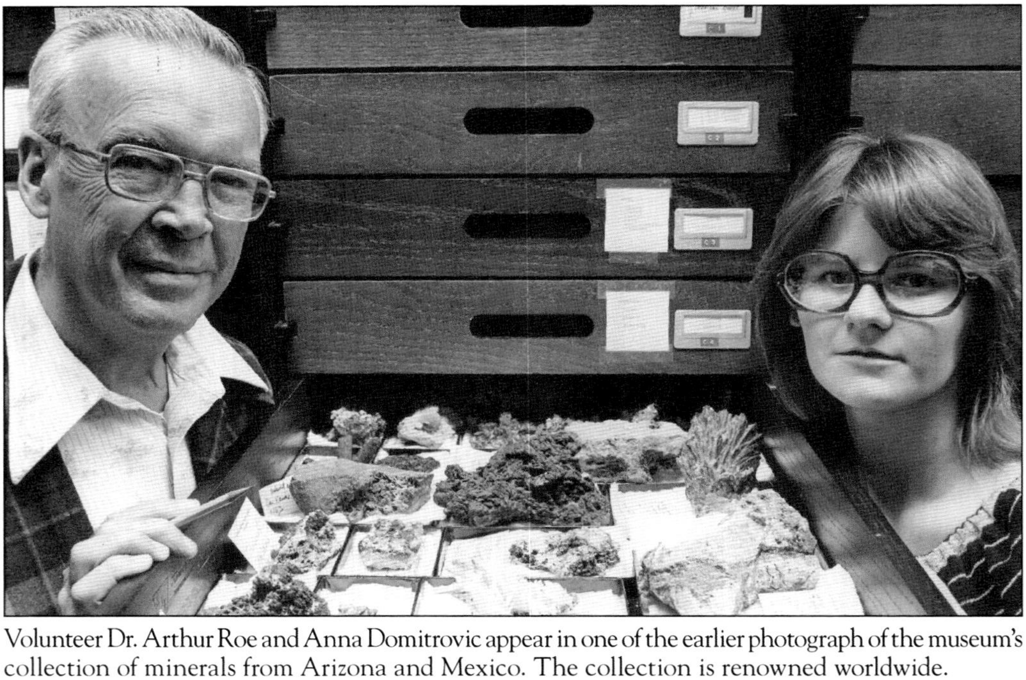

Volunteer Dr. Arthur Roe and Anna Domitrovic appear in one of the earlier photograph of the museum's collection of minerals from Arizona and Mexico. The collection is renowned worldwide.

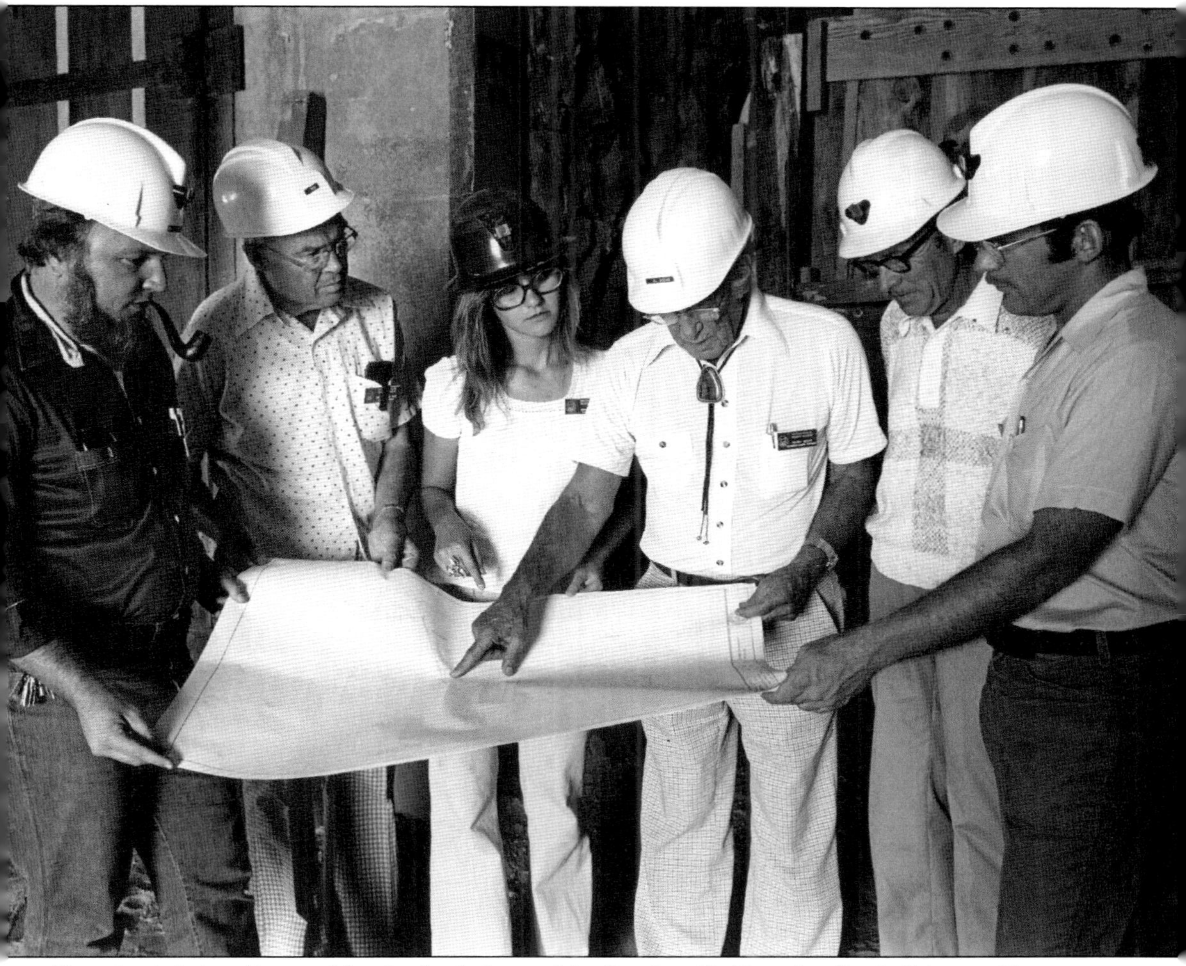

Planning is being discussed for the final phase of the Stephen H. Congdon Earth Sciences Center on June 5, 1978. Seen from left to right are curator Bill Panczner, Arthur Roe, Anna Domitrovic, Manny Hecht, director Dan Davis, and Peter Kresan.

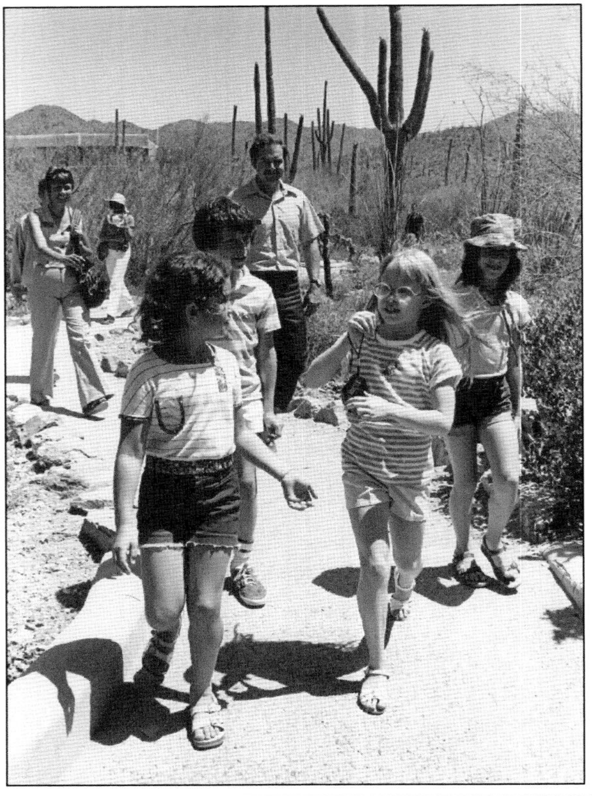

Amy Carter (right front), daughter of Pres. Jimmy Carter, is seen here among friends enjoying a day at the Arizona-Sonora Desert Museum. Director Holt Bodinson and Doris Ready follow close behind.

Sen. Barry Goldwater (with camera) toured the old animal line of enclosures with Director Larson in the 1970s. Senator Goldwater was honored for his support of the museum with the dedication to him of the mountain lion portion of the Mountain Habitat.

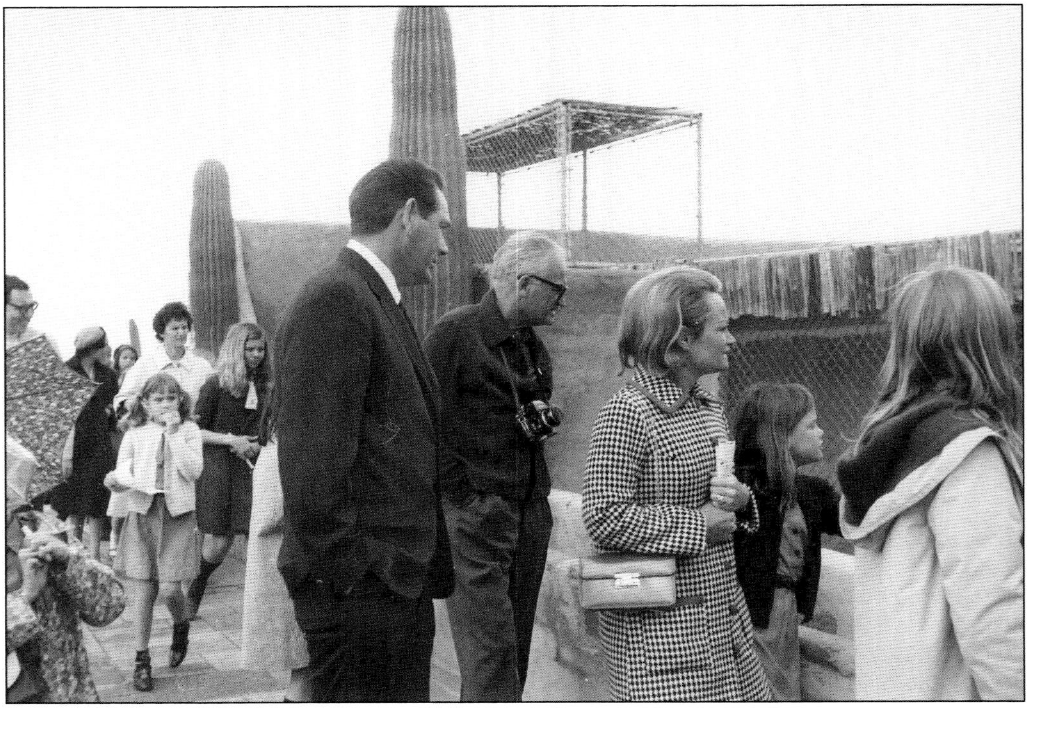

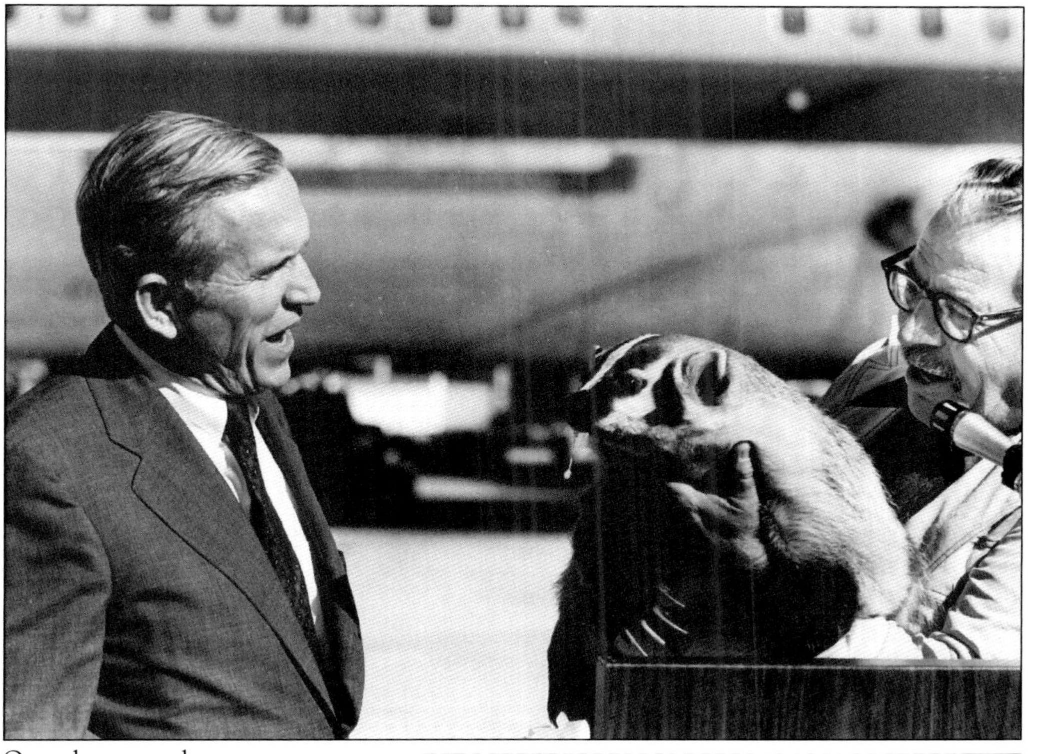

Over the years, the museum has been host to many notable personalities. Here Hal Gras introduces Frank Borman, former astronaut and president of Eastern Airlines, to B. B. Badger, a member of the Desert Ark. In his teenage years, Borman attended Tucson High School, where a badger was the school's mascot.

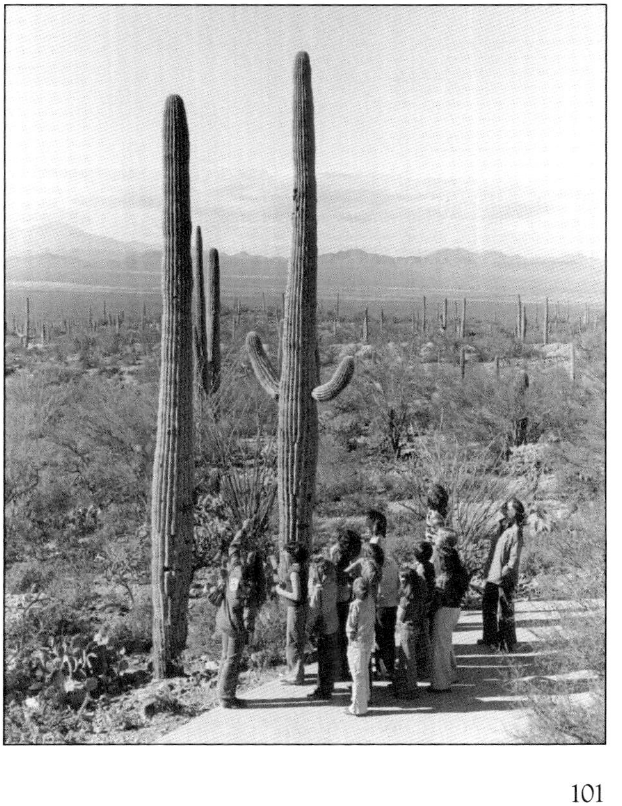

Since the docent program was instituted in 1972, as specialized volunteers the docents have provided invaluable service to the museum. They started by touring school or other groups on grounds, such as this docent is doing. Soon their duties expanded greatly, and currently 200 docents volunteer to conduct a wide variety of museum activities.

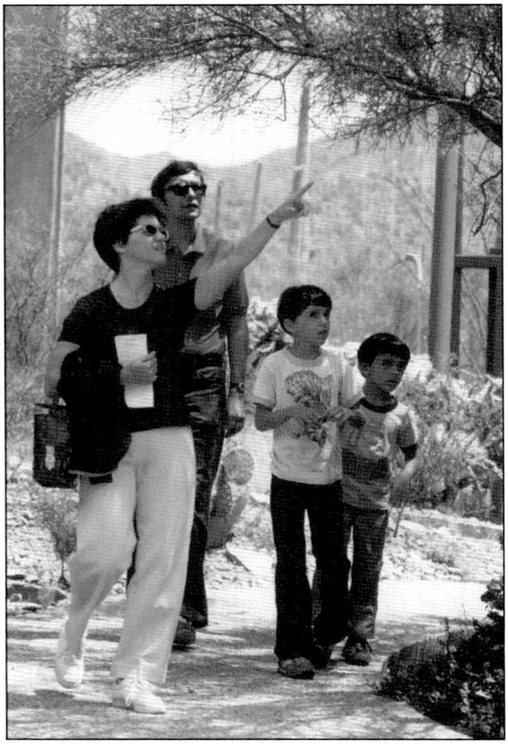

The popular walk-through aviary built in 1979 replaced a similar smaller one and was built with funds provided by the Stonewall Foundation. This aviary contains numerous species of plants and over 40 species of free-flying birds indigenous to the Sonoran Desert. Photographers find this exhibit user-friendly for capturing high-definition images of a variety of birds.

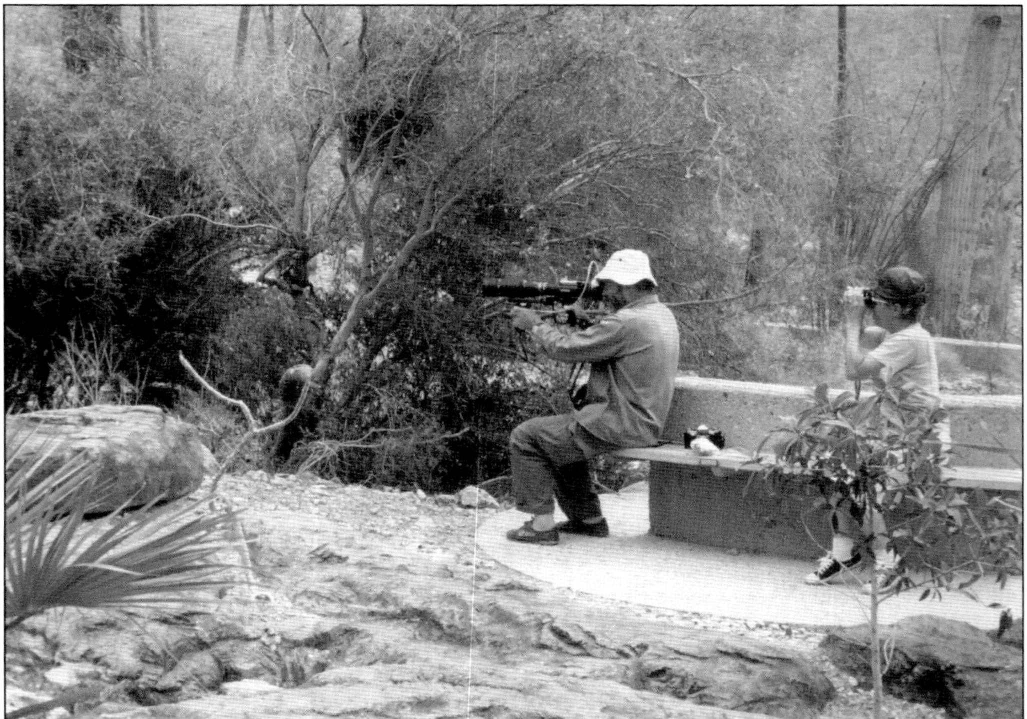

Four

1980s

The 1980s were a decade of transition for the Arizona-Sonora Desert Museum. During that time period, a comprehensive plan for expansion was put into action. The completion of the Earth Sciences Center in 1981 brought to a successful conclusion the museum's plans to interpret the geological history of the Earth. When the popular Arizona-Sonora Desert Museum–sponsored television series *Desert Trails*, which ran for 32 years, went off the air in 1985, it was soon replaced by another series, called *The Desert Speaks*, on the CBS station KOLD–TV. Later the program was coproduced by the museum and Arizona Public Media.

In the 1980s, the last early employees left the museum. Curator of small mammals Merritt Keasey departed, and longtime animal interpreters Hal and Natie Gras retired in 1985. Hal Gras was the last of the original staff members employed at the museum. The founder of the Desert Museum, Bill Carr, passed away that same year, closing a chapter in the early history of the museum.

A new chapter emerged for animals in captivity at the Arizona-Sonora Desert Museum, including the birth of four Mexican gray wolves, four jaguarundis, one jaguar, two masked bobwhite quail, one bighorn lamb, and two third-generation captive-born ocelots. Captive breeding of three out of four parrot species, along with a rare hatching of caracaras, made the 1980s a productive decade for animal life.

New animal enclosures were created through an aggressive capital campaign. The Mountain Woodland Habitat was established to provide more space for the mountain lion and bear enclosures. A hummingbird aviary opened October 12, 1988, with eight of the hummingbird species endemic to the Sonoran Desert Region displayed. The exhibit was made possible through funding by Mr. and Mrs. Sidney F. Tyler. On March 26, 1989, a renovated Tunnel called Life Underground was dedicated to William H. Carr, who pioneered the original exhibit.

By the end of the decade, the Arizona-Sonora Desert Museum was renowned in stature as one of the premier institutions of its type in the world. *Parade* magazine proclaimed the Arizona-Sonora Desert Museum one of the 10 best zoos in the nation. That same year, the museum had over 600,000 visitors in attendance and Dan Davis resigned as its director.

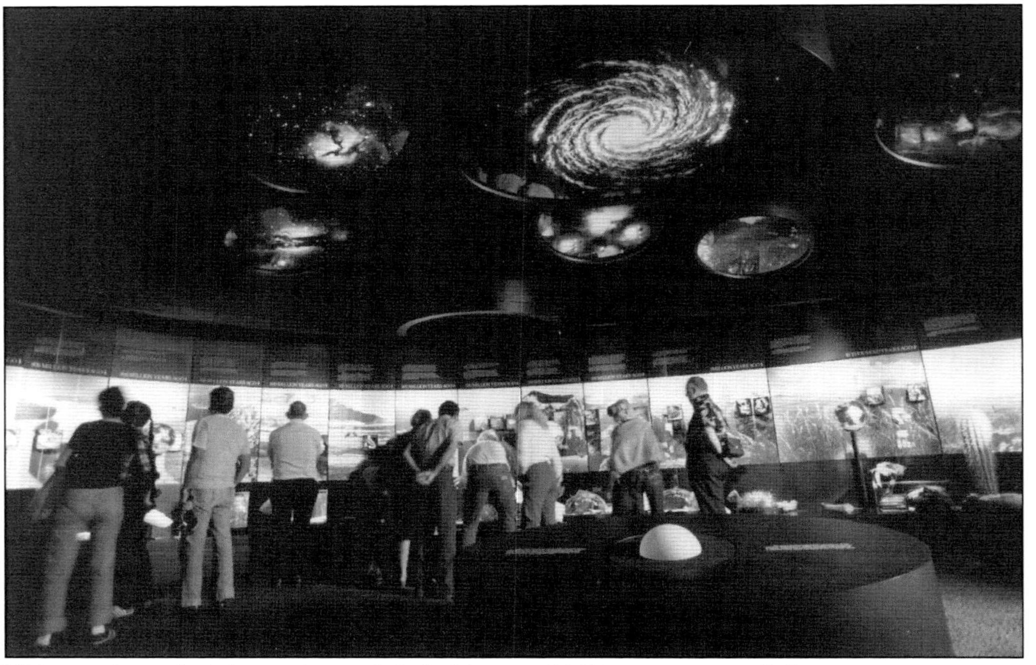

Visitors gain a celestial experience while looking at a 5 billion-year pictorial dioramic representation depicting the history of the Sonoran Desert. Paintings of the Milky Way, along with multiple photographs of nebulae, adorn the ceiling of the exhibit hall in the Earth Sciences Center. Funded by a donation from Peggy Speed, a moon rock consisting of a two-inch cube of basalt on loan from NASA was acquired in 1992. The special rock augments the exhibit as part of the origin of the Earth and Moon interpretation. Speed was a trustee, and she and her husband William were longtime supporters of the museum.

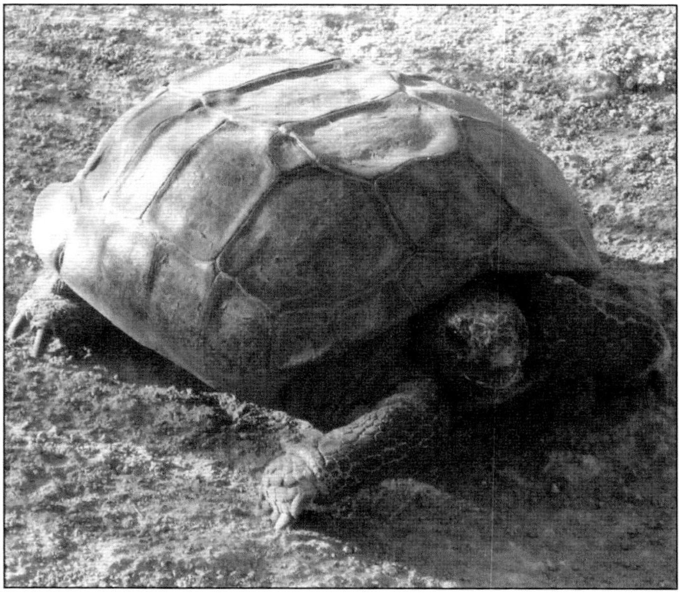

Founded in 1984, the Arizona-Sonora Desert Museum Tortoise Adoption program, sanctioned by the Arizona Game and Fish Department, has been instrumental in finding homes for displaced desert tortoises. Many have been negatively impacted by human development, and the program has been instrumental in locating people who are willing to adopt and provide proper care for these long-lived (over 80 years) reptiles. The tortoises remain the property of the state. (Photograph by William Ascarza.)

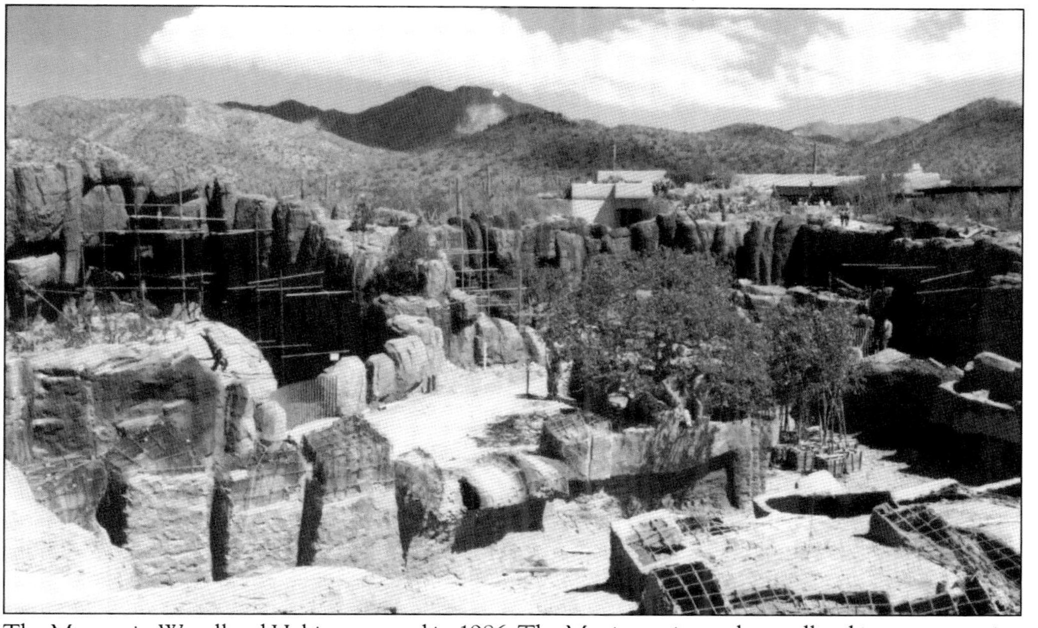

The Mountain Woodland Habitat opened in 1986. The Mexican pine-oak woodland is representative of the flora and fauna found in the Sky Islands of southeastern Arizona, which rise above the desert. Exhibits include mountain lions, thick-billed parrots, Mexican gray wolves, white-tailed deer, and black bears. Trees include cottonwood, juniper, pine, and oak.

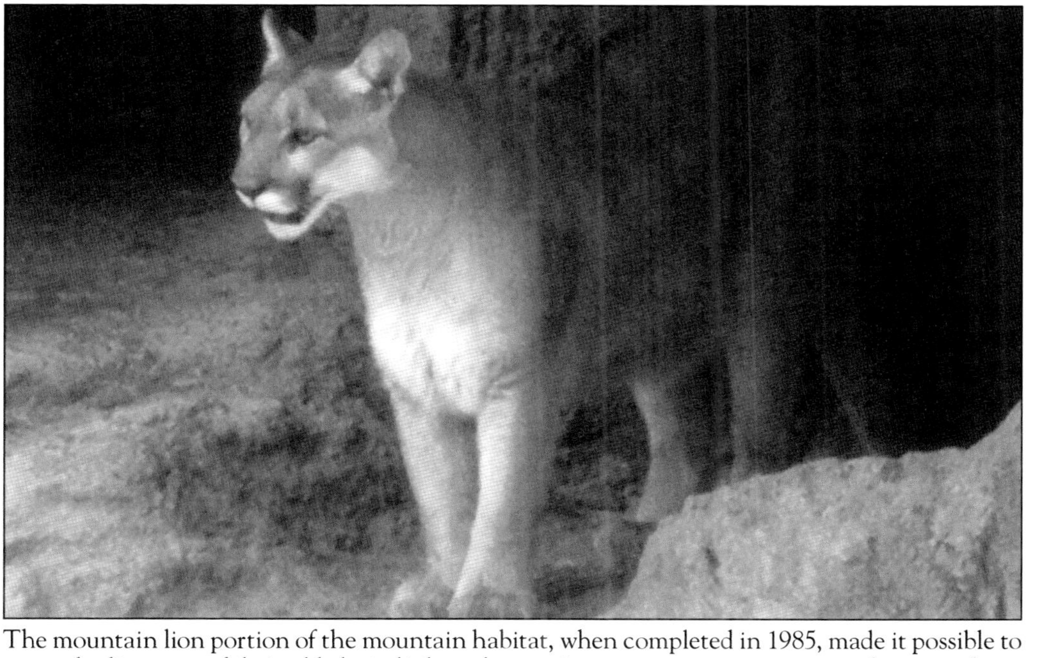

The mountain lion portion of the mountain habitat, when completed in 1985, made it possible to move the lions out of their old chain link enclosures into a new, spacious, manmade, naturalistic enclosure. A resting area with concealed heating for cold weather and a hidden evaporative cooler for summer heat lured the lions close to a window in one viewing area. Even a stream-like swimming area was built into the enclosure. (Photograph by William Ascarza.)

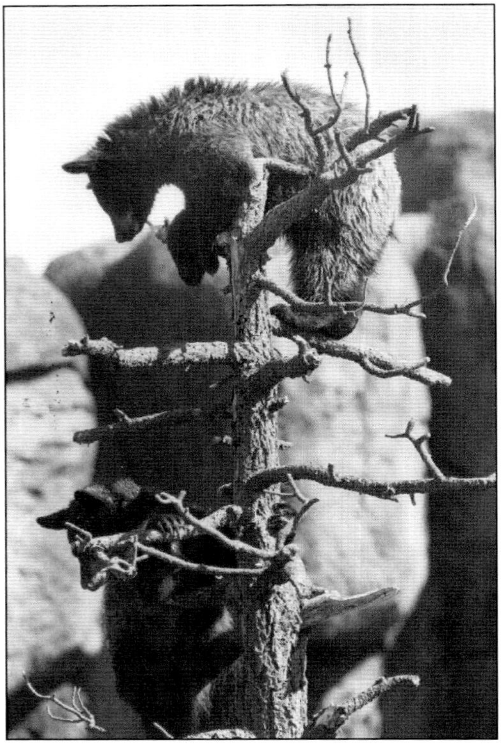

Two black bear cubs from Montana are shown in 1986 enjoying their new enclosure, which mimics their natural surroundings in the mountain woodland environment. Both black bears remained active through 2007, when they were officially retired and placed in geriatric quarters. Black bears are residents of several of the high mountain ranges that surround Tucson. They remain active throughout the year, foraging for wild berries, roots, and young plants. (Both photographs by Tim Fuller.)

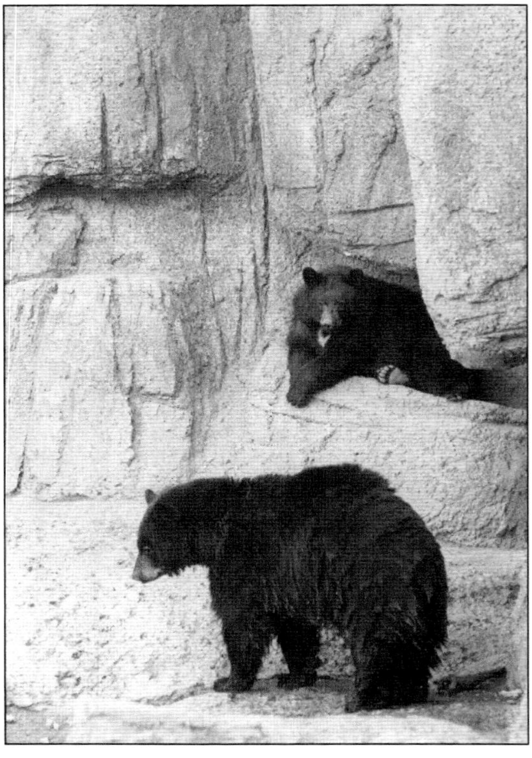

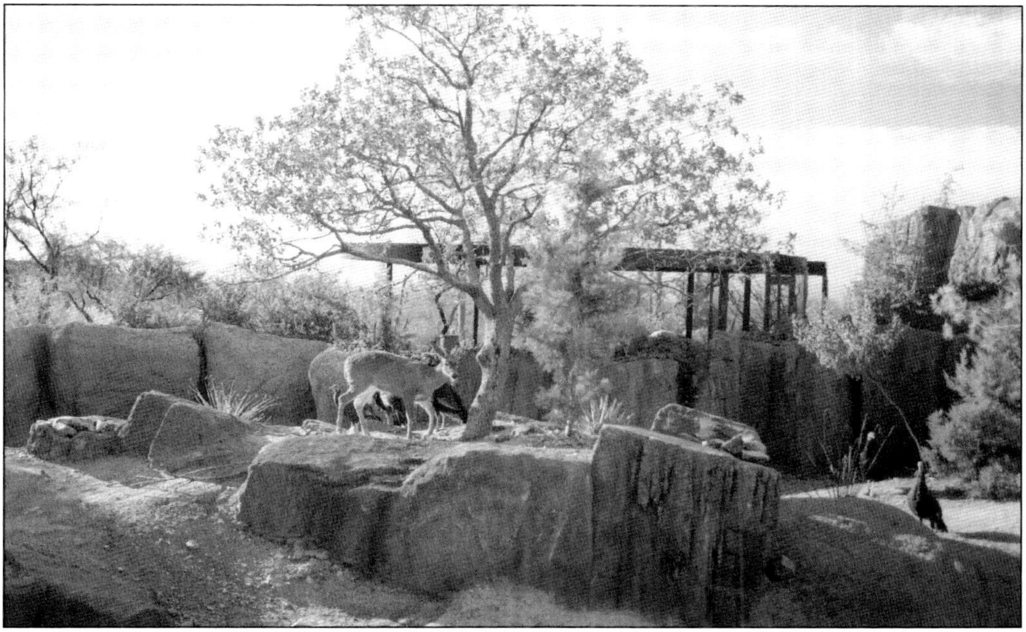

White-tailed deer and wild turkey also inhabit the mountain woodland. The white-tailed deer is an herbivore commonly found in the foothills of mountain ranges in southeastern Arizona. The wild turkey, although considered endangered in the region, has recently been reintroduced with success to the Sky Islands of southeastern Arizona. (Photograph by Tim Fuller.)

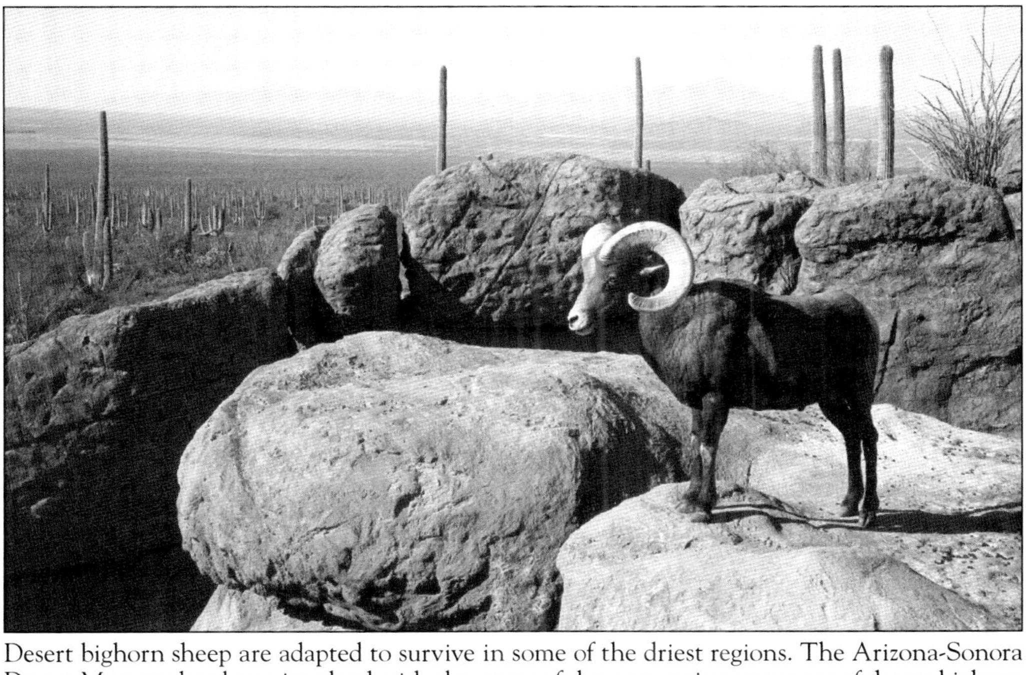

Desert bighorn sheep are adapted to survive in some of the driest regions. The Arizona-Sonora Desert Museum has been involved with the successful conservation measures of desert bighorn sheep within Ironwood Forest National Monument located near the Silverbell Mountains 30 miles northwest of Tucson. (Photograph by Robert Perrill.)

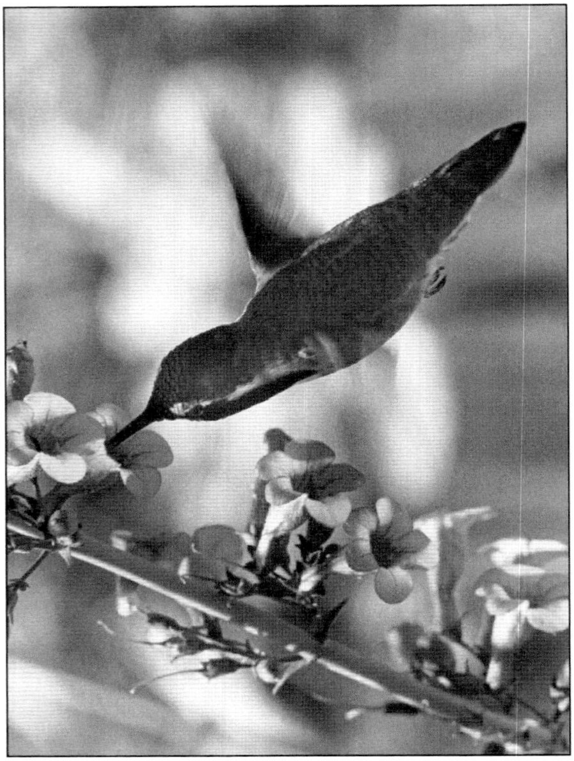

A hummingbird is gathering nectar from a flower in the hummingbird exhibit. Built in 1988, the hummingbird aviary is a popular exhibit for visitors to view some of the 17 species of hummingbirds found in Arizona out of over 300 species of hummingbirds found in the Americas. (Photograph by Al Morgan.)

On October 24, 1989, director emeritus William H. Woodin (left), honorary cofounder Phoebe Pack, and board of trustees vice president Bernard W. Simons Jr. prepare to cut the ribbon on the renovated Tunnel, renamed Life Underground and dedicated to museum cofounder William H. Carr.

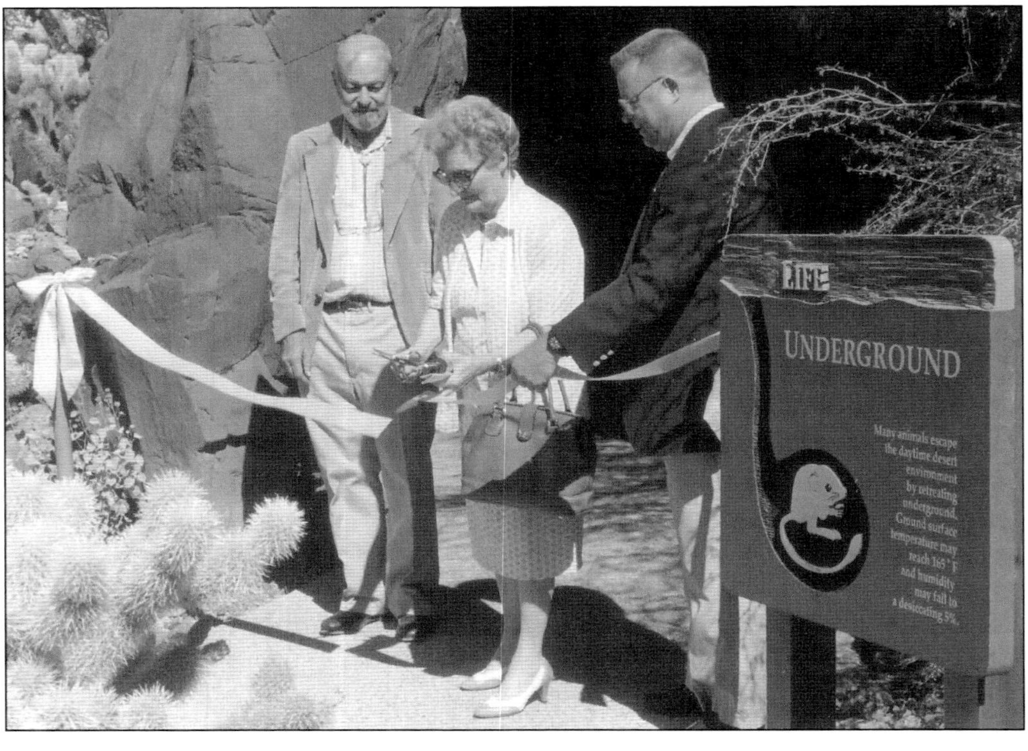

Five

1990s

The 1990s were a decade of refinement in programming, exhibition, and staffing at the Arizona-Sonora Desert Museum. Formerly director of Seattle's Woodland Park Zoo from 1976 to 1984, David Hancocks assumed the directorship of the museum on January 1, 1990. He was involved in the planning and implementation of three new exhibits on grounds, including the grasslands, desert foothills, and riparian areas. Throughout the decade, the museum gained additional exposure through national television, educational programs, new interpretation sites along with updated animal enclosures, and publications. Rick Daley assumed the role of director of the museum in June 1998.

In 1991, *The Desert Speaks* went national. The program is produced by KUAT-TV, channel six, Tucson's PBS affiliate, and is underwritten by ASARCO, Inc. ASARCO, Inc. is a longtime friend of the museum with major contributions toward the construction of the Earth Sciences Center and other generous donations. Noted actor Don Collier and Lauray Yule cohosted the show, which emphasized "plants, animals, ecology and geology of the Sonoran Desert Region." Members of the Arizona-Sonora Desert Museum staff and representatives from other environmental institutions often appeared as guests on the program.

A publications department, the ASDM Press, was established in 1994 under the direction of Steven J. Phillips, who served as its manager. Publications throughout the decade included the *Secret Lives of Hummingbirds*, *Wild Food of the Sonoran Desert*, *Tucson Mountains Trail Guide*, *Mount Lemon Road Guide*, and *The Natural History of the Sonoran Desert*.

The number of education programs increased in 1992, including the highly successful Amigos del Desierto program, with the museum entering annually into a special educational partnership with an elementary school having a large Hispanic enrollment. A successful volunteer program was established that expanded rapidly, and now over 300 volunteers assist in 60 different jobs.

By the close of the decade, new pollinator gardens had been established, featuring 225 species of flowering plants for pollinating bees, bats, butterflies, and hummingbirds. A Migratory Pollinator Project began, consisting of a five-year study to monitor four migratory pollinator species along the nectar corridors of western Mexico and the southwestern United States. Tracking of white-winged doves, lesser long-nosed bats, rufous hummingbirds, and monarch butterflies moving through the Sonoran Desert was undertaken. A new javelina enclosure was created using 20,000 square feet of Invisinet, depicting their natural environment.

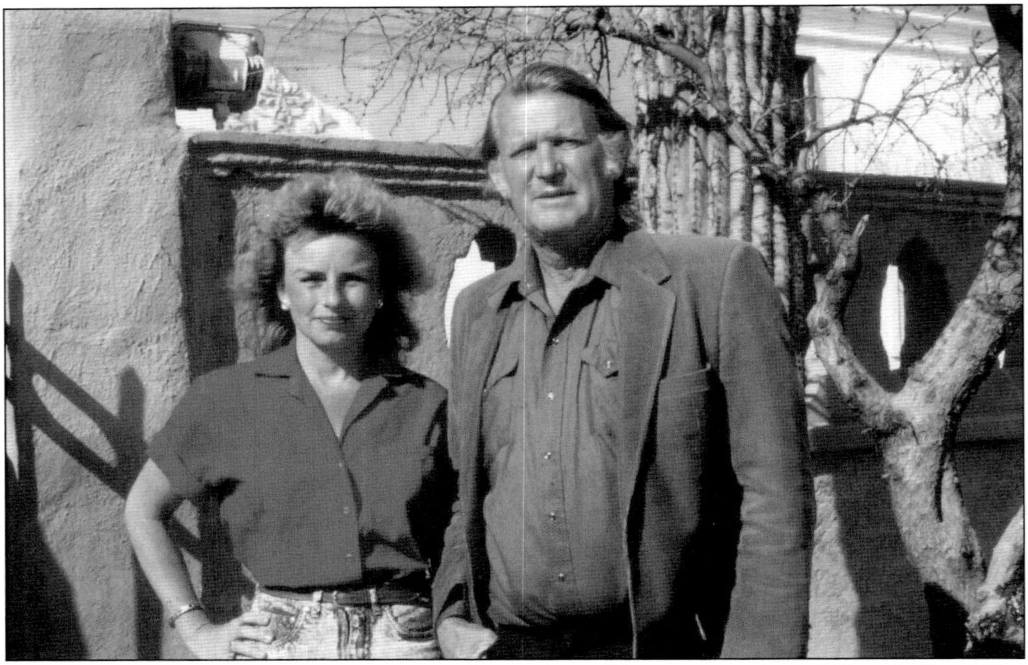

Funded by ASARCO, Inc., *The Desert Speaks* was hosted by Lauray Yule and Don Collier. This popular television show is produced in collaboration with the Desert Museum, The Nature Conservancy (Arizona), and KUAT-TV.

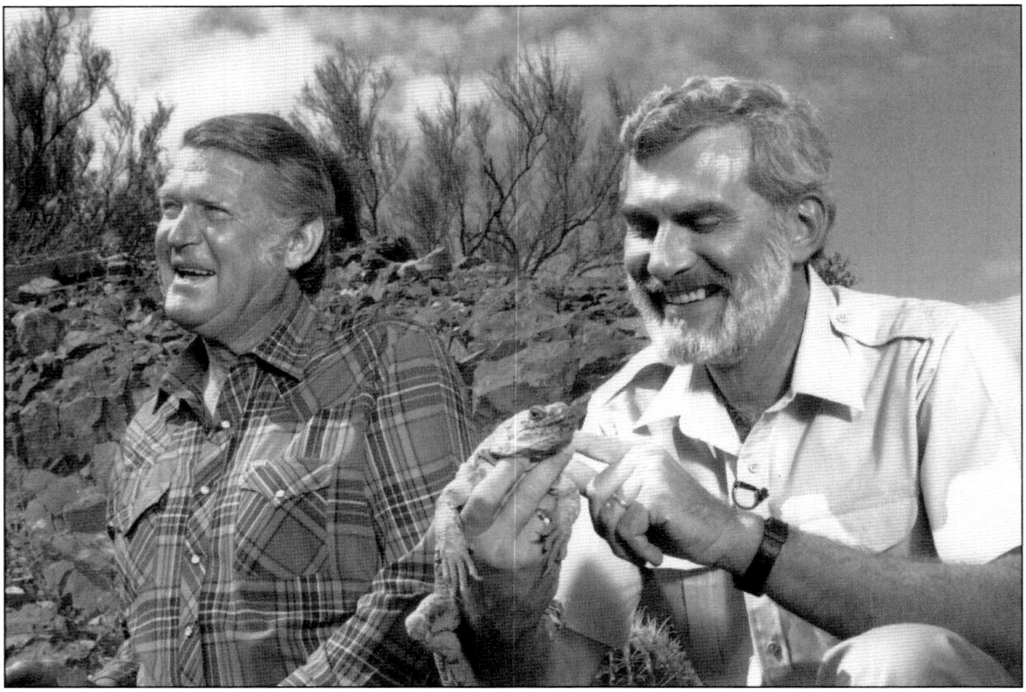

Don Collier (left) and Desert Museum curator Howard Lawler present a San Esteban chuckwalla in front of the camera during the filming of the *The Desert Speaks*.

Seen in the Arizona-Sonora Desert Museum's representation of the grassland habitat found in central and southeastern Arizona, soaptree yucca is a signature plant of the region. Also known as Spanish bayonet, its trunk can reach 23 feet high, and its creamy white flowers appear in May and June on the upper half of its stalk reaching up to 10 feet tall. (Photograph by William Ascarza.)

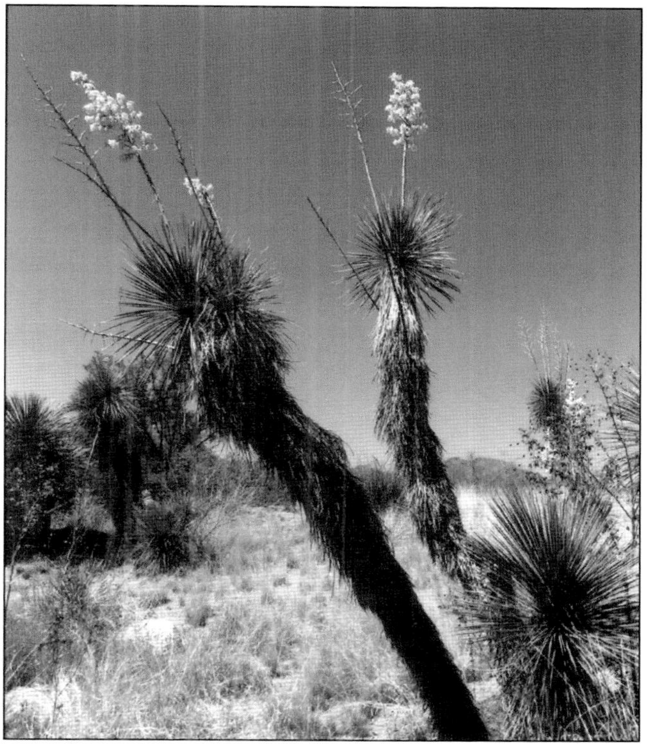

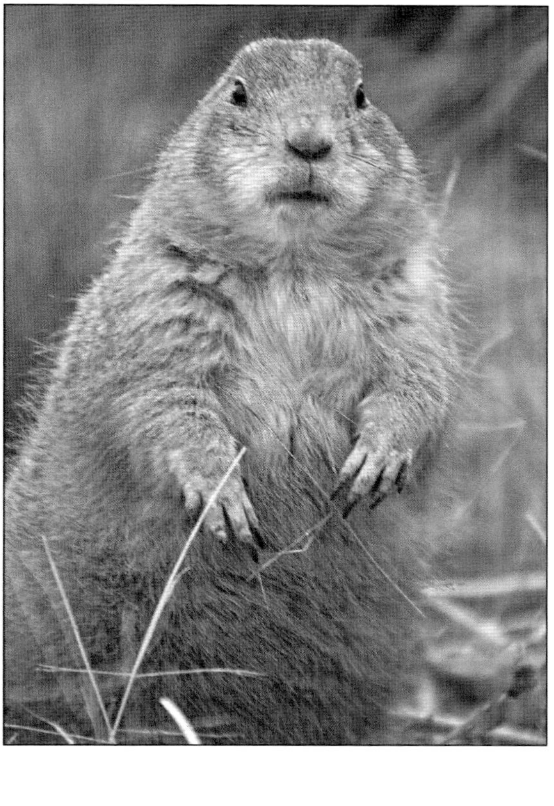

The black-tailed prairie dog was once a common inhabitant of the grasslands in southeastern Arizona. Considered anathema among ranchers, this species was extirpated in southern Arizona until the Arizona Game and Fish Department began reintroducing black-tailed prairie dogs into the Las Cienegas Natural Conservation Area near Sonoita in southeastern Arizona in 2008. The grassland exhibit at the Arizona-Sonora Desert Museum established in 1992 includes this animal, which was once endemic to the region. (Photograph by Kenny Don.)

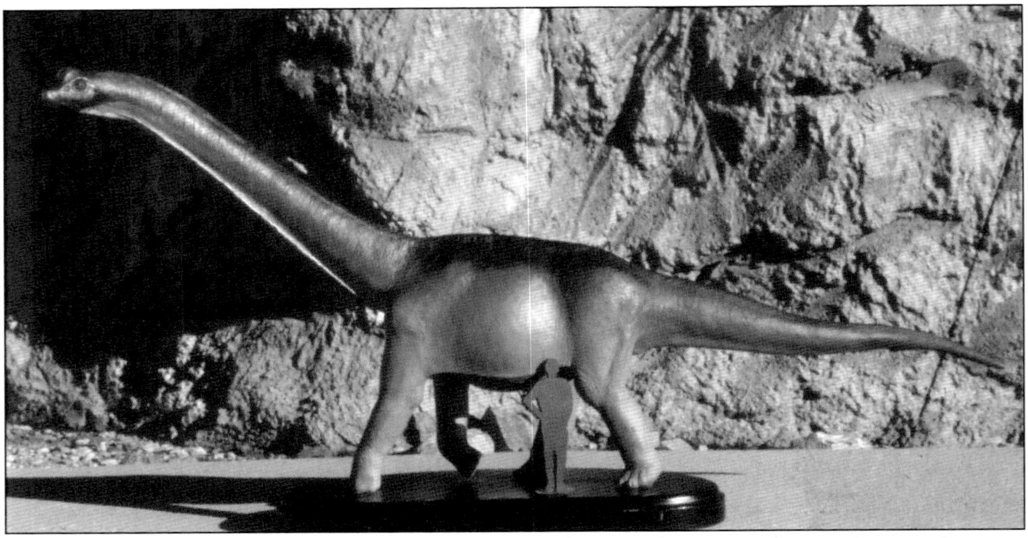

In November 1994, amateur fossil hunters discovered the partial remains of a rare Arizona dinosaur called sonorasaurus, a type of brachiosaur, embedded in sandstone. Sonorasaurus or *Sonorasaurus thompsoni* was named after its discoverer, Richard Thompson, and the Sonoran Desert region in which it was discovered. It is believed to have measured 50 feet long and 20 feet tall, weighing between 20 and 30 tons.

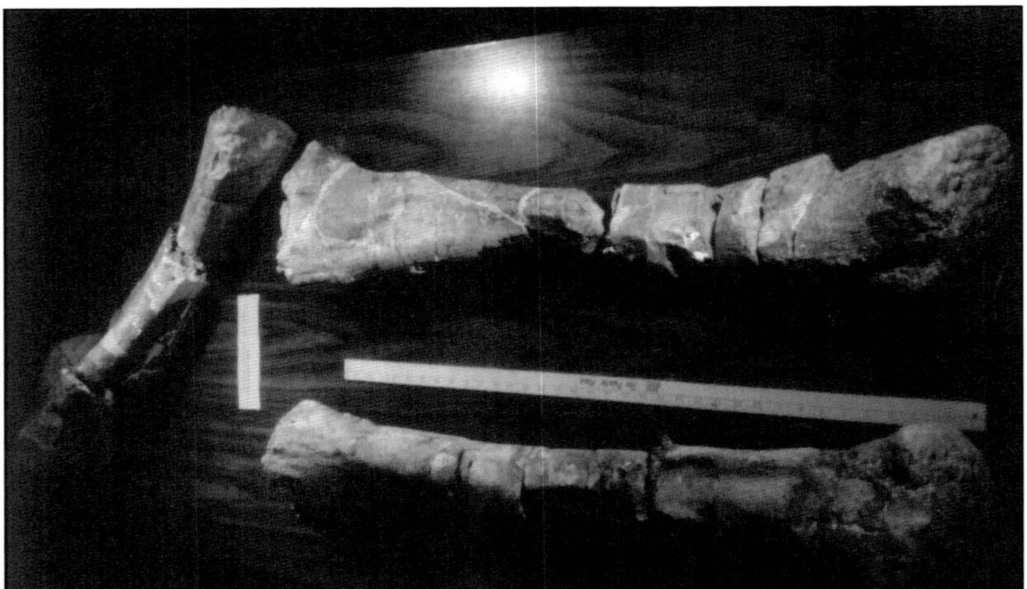

A section of a fossilized leg bone was the first evidence discovered of the 100-million-year-old sonorasaurus. Additional remains discovered included hip and rib fragments, part of a shoulder blade, and the rounded surface of a limb joint. Over the course of five years, all the visible bones, encompassing 30 percent of the skeleton, were removed and shipped to the Desert Museum's Earth Science Department, where they are currently in storage.

The Desert Museum was entrusted with the task of conducting the principal excavation of sonorasaurus at the remote site on state trust land in the Whetstone Mountains. The excavation team, under the direction of project manager and museum paleontologist Ron Ratkevich, used a mixture of hammers, chisels, and a tool called a Pneumatic Air Scribe to remove the bones from their sandstone matrix.

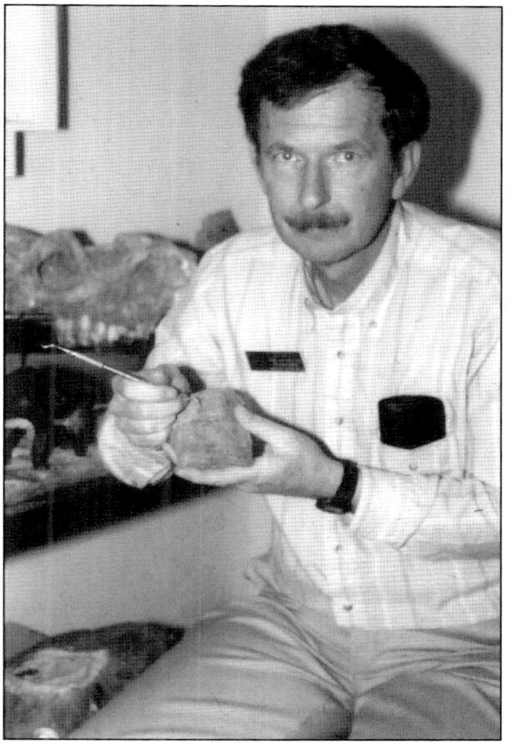

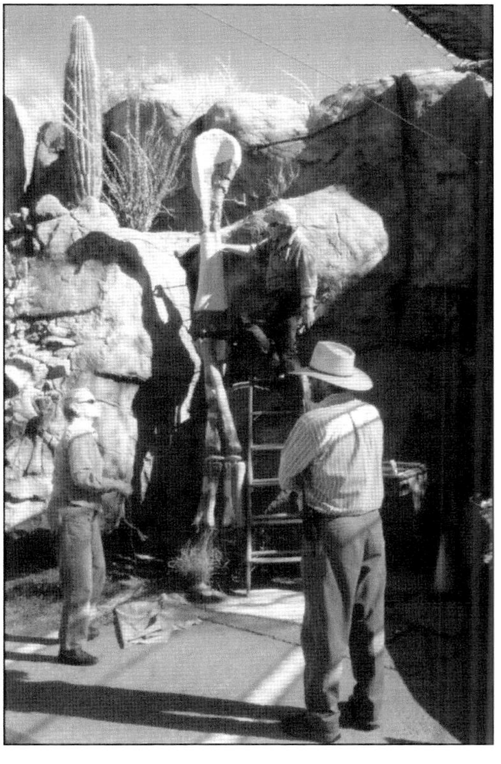

Detailed castings of dinosaur bones are seen here being assembled for the Ancient Arizona exhibit, which opened in 2000. Another portion of the outdoor exhibit includes a hands-on children's fossil dig in sand and also a cliff face representing a replica of the excavation site with sonorasaurus bones in place.

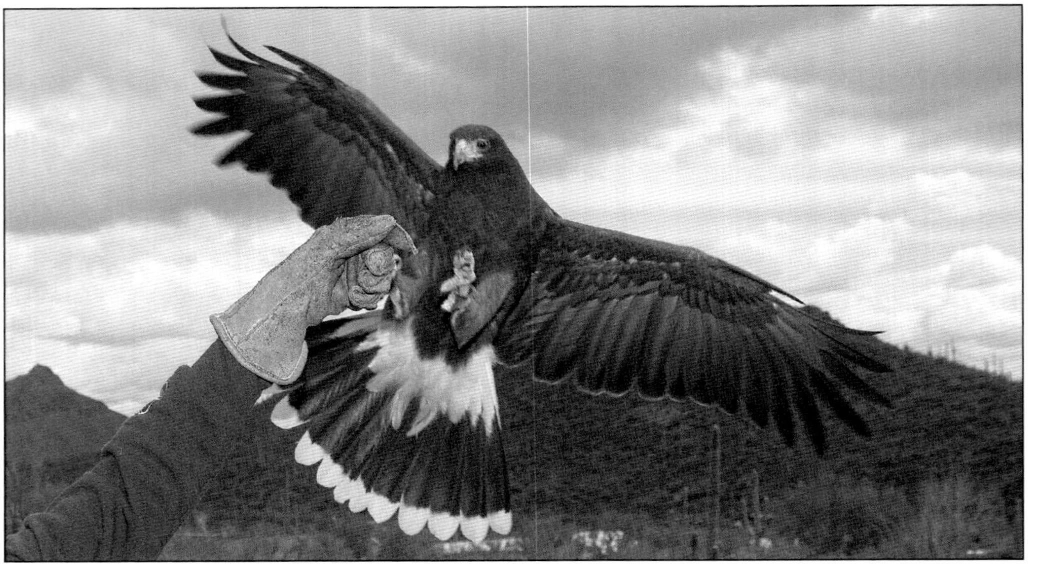

The Raptor Free Flight program debuted in 1996. Because of the intense summer heat, the program runs during cooler months, taking place twice a day from late October to early April. This incredible program allows visitors the opportunity to witness the free-flying natural activities of Sonoran Desert birds of prey in a natural setting. The raptor seen here is a Harris's hawk, which is landing on the glove of the trainer. (Both photographs by Kenny Don.)

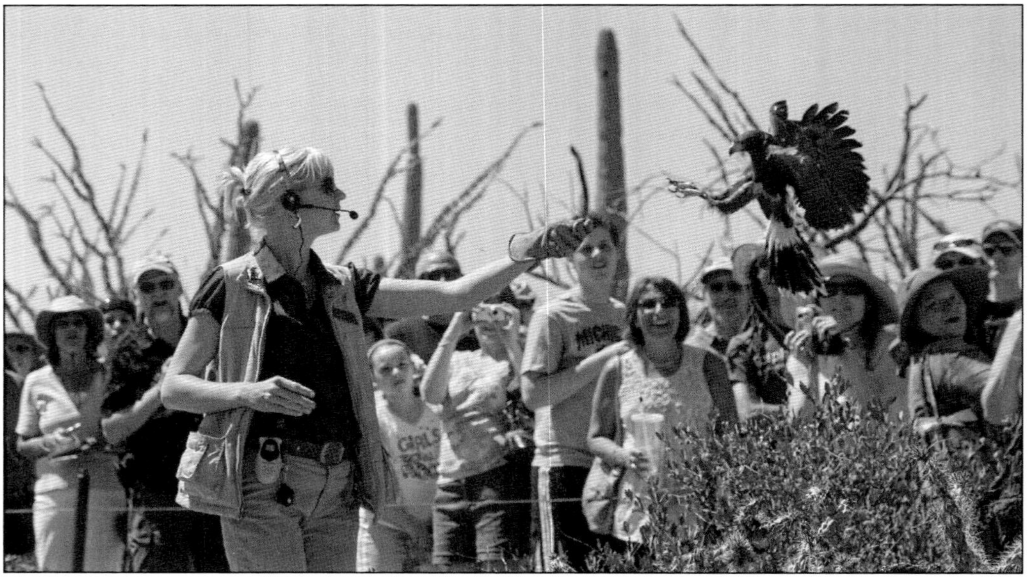

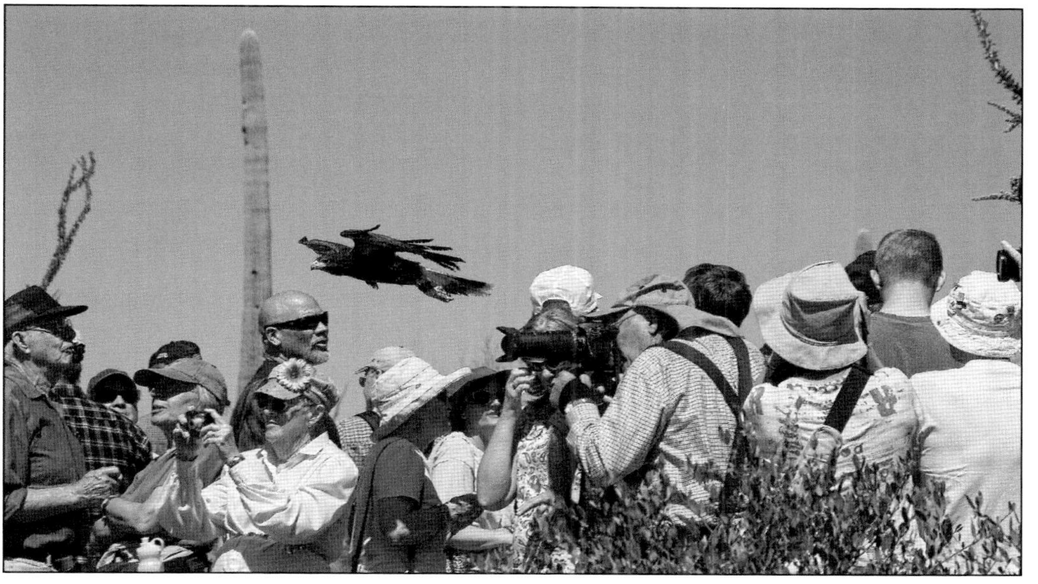

A Harris's hawk cuts through the air above an astonished and amazed crowd. Individual species of birds presented include barn owls, great horned owls, ferriginous hawks, gray hawks, and prairie falcons. Two species of birds, Chihuahuan ravens and greater roadrunners, although not raptors, are also presented. (Both photographs by Kenny Don.)

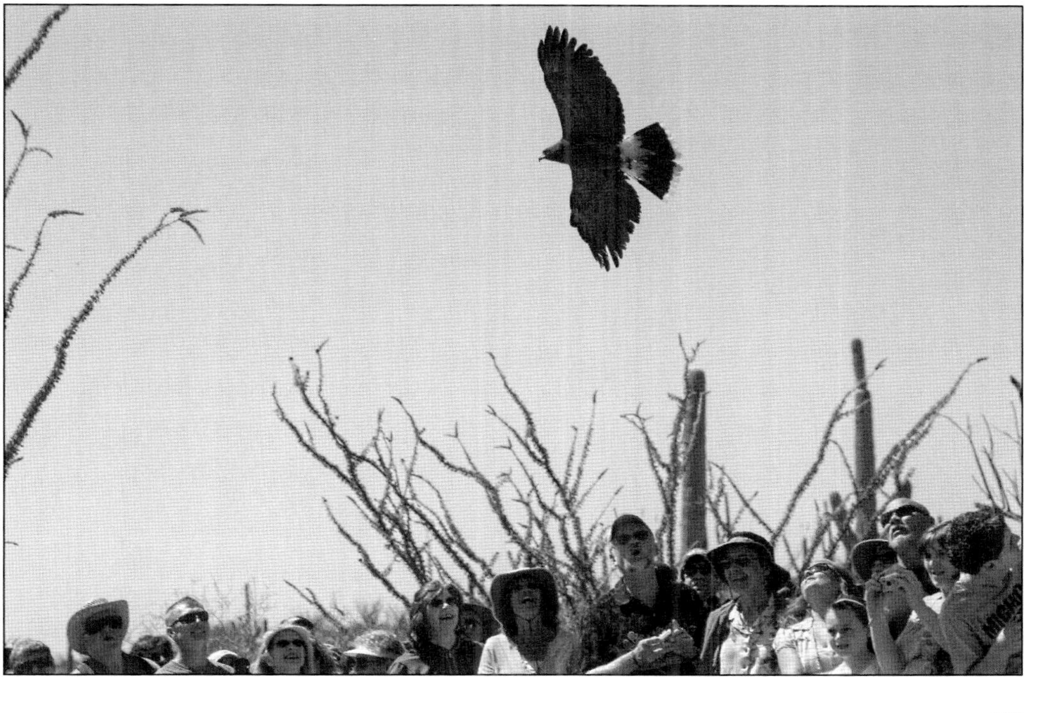

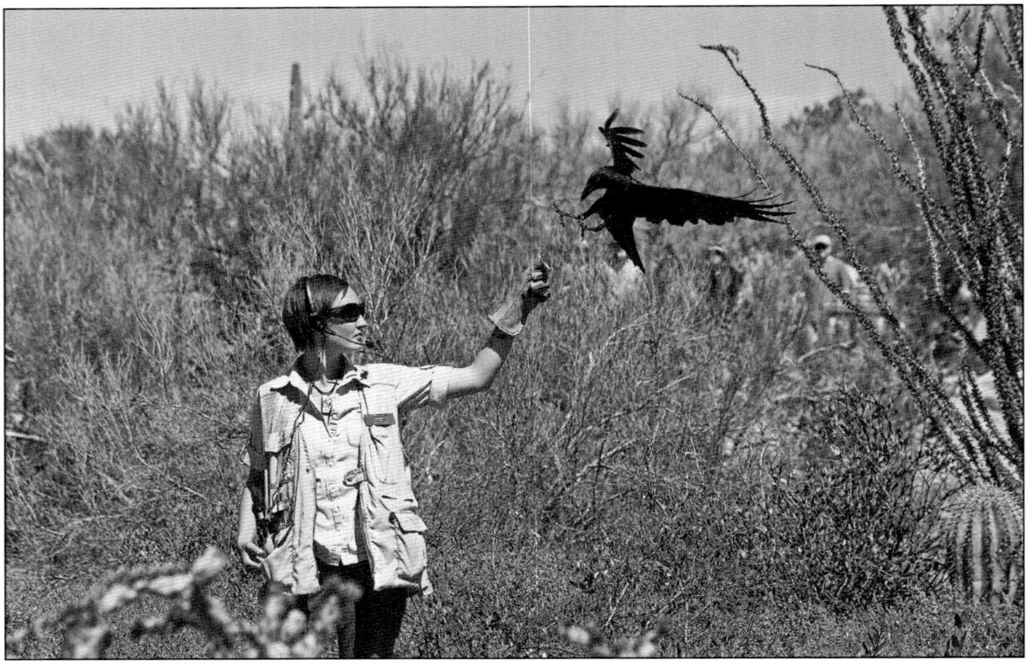

A Chihuahuan raven is seen interacting with a raptor trainer during the Raptor Free Flight. The American Zoo and Aquarium Association commented on the uniqueness of the Raptor Free Flight program and equated it to the visitor taking a tour through the open desert with a raptor biologist explaining the natural behavior of the bird flying overhead. (Both photographs by Kenny Don.)

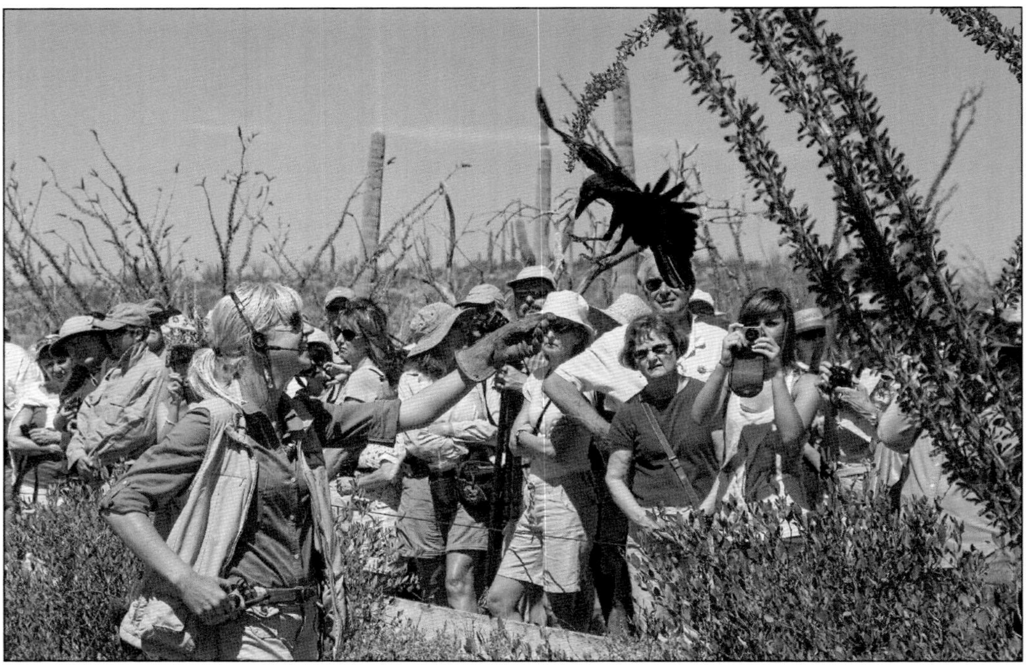

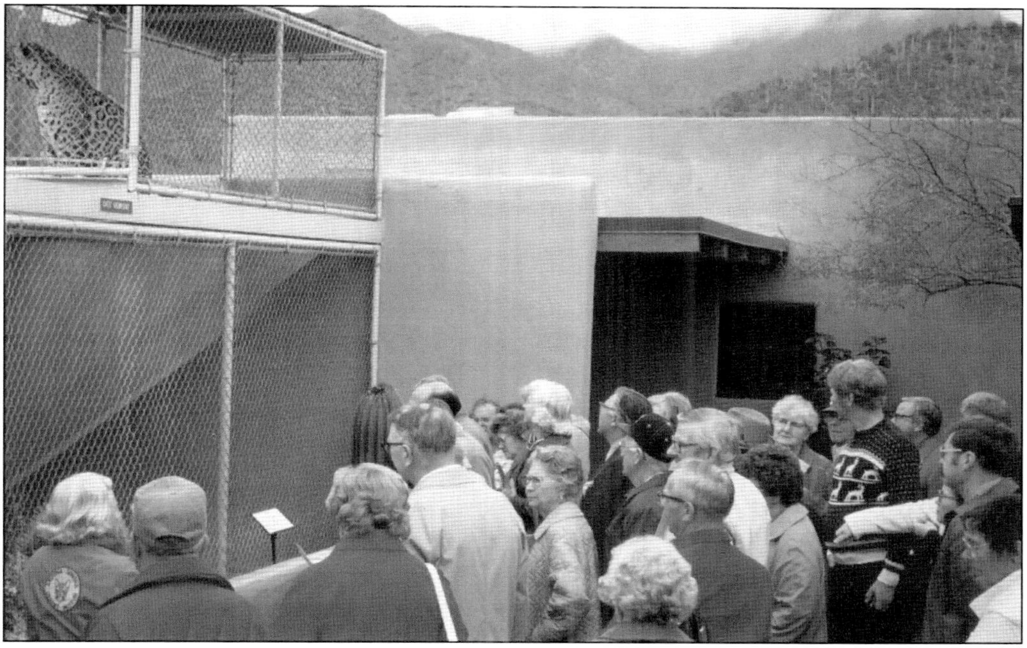

The jaguar was a real eye-catcher among visitors in the 1990s. These beautiful cats are a rare sight in southern Arizona. Even more so is the old enclosure shown here with the animal's upper walkway. This enclosure no longer exists, and a new more naturalistic one with larger space is being planned for the future.

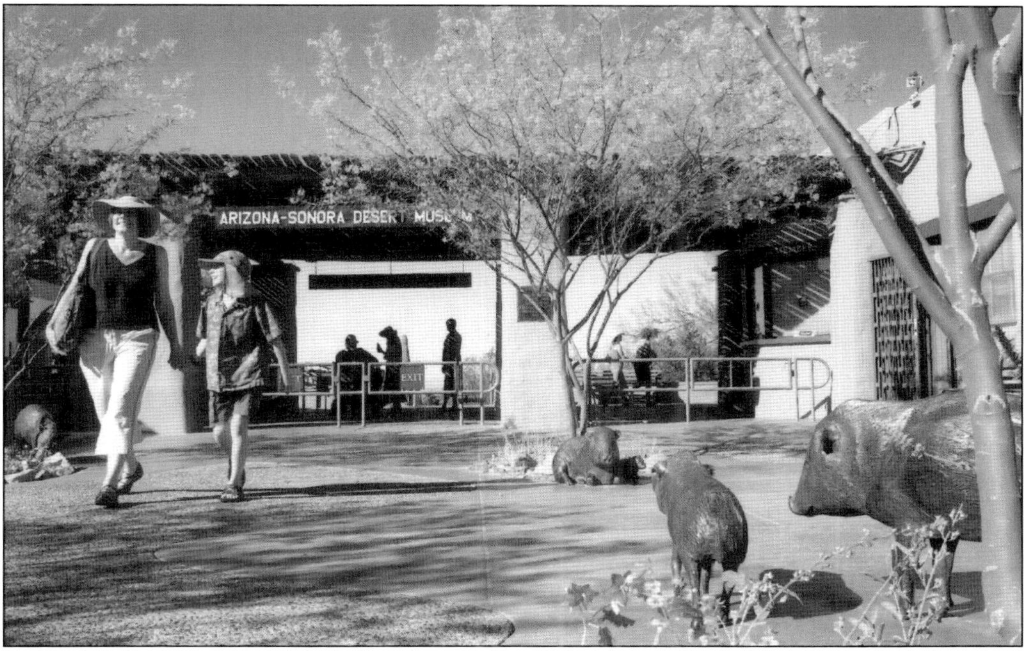

The front entrance at the Arizona-Sonora Desert Museum is shown as it currently appears. The annual attendance of the museum today is nearly half a million visitors. (Photograph by Rhonda Spencer.)

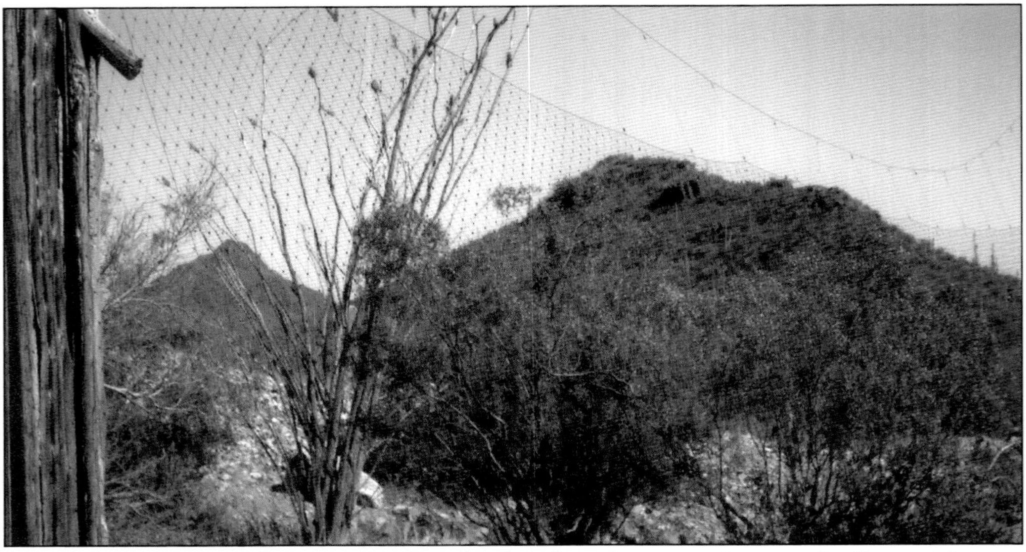

Already innovative in the creation of artificial rock and open enclosures, the Desert Museum expanded its proficiency in creating animal exhibits by the invention of Invisinet. Developed by museum director of design and planning Ken Stockton, Invisinet has helped provide naturalistic authenticity for animal enclosures, including those of the javelina and coyote found on the half-mile Desert Loop Trail. Visitors have the benefit of viewing animals through the Invisinet fine stainless steel netting, which is nearly invisible to the eye and the camera. (Photograph by William Ascarza.)

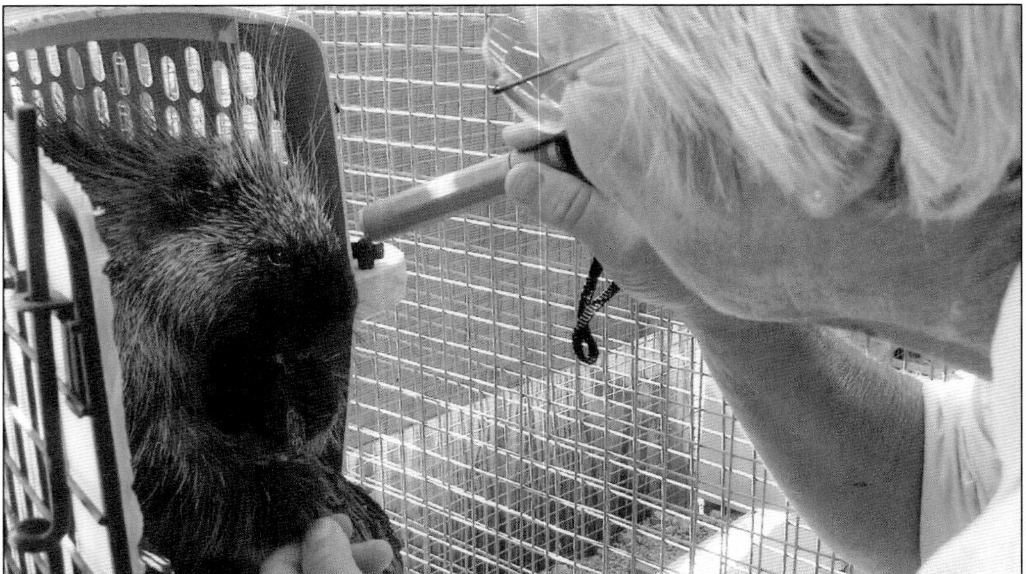

Two veterinarians make regular visits to the museum to care for the animals. Members of the museum's staff conduct various enrichment and training sessions with the mammals, such as the mountain lions, to expedite their care by the veterinarians. Here Dr. Sophia Kaluzniacki examines a porcupine. Dr. James Jarchow is a veterinarian for reptiles and other non-mammals. (Courtesy of Sloane Haywood.)

Six

2000s

In the new millennium, the Desert Museum continues to expand its outreach through education, exhibition, conservation, recreation, and scientific inquiry. Robert J. Edison assumed the position of executive director during the decade.

In this decade, the ASDM Art Institute opened to provide an additional venue for students to artistically interpret the flora and fauna of the Sonoran Desert. The Baldwin Education Building and the Warden Oasis Theater opened, providing venues for interpretive programs, presentations, meetings, and social events. A major exhibit, Life on the Rocks, opened in 2007. It depicts wildlife, plant life, and minerals found both on the surface and underground in the rocky slopes of the Tucson Mountains.

Technology became a museum focus with the creation of the ASDM Digital Library. The online Sea of Cortez faunal database and the ASDM digital service features an extensive, searchable, multimedia, Web-based catalog of the plants, animals, minerals, landforms, and landscapes of the Sonoran Desert Region.

Several enhancements in educational programs occurred through this decade. The first annual Laurel Clark Earth Camps began in June 2005, cosponsored by the Arizona-Sonora Desert Museum and University of Arizona. Named to honor a *Columbia* space shuttle astronaut, these camps foster among youth a greater understanding of the desert and other ecosystems worldwide.

New live animal demonstrations were developed. Live and (sort of) on the Loose features live venomous reptiles and interprets their natural history, dispels myths, and presents truths about Gila monsters, Mexican beaded lizards, and rattlesnakes. The Running Wild program presents a porcupine, ringtail, pelican, macaws, and other animals exhibiting natural behaviors. Raptor Free Flight, which originated in the 1990s, features impressive free flights of raptors in the open desert.

The museum continued to refine its volunteer opportunities, including participation in the Invaders of the Sonoran Desert Region program, created in response to concern pertaining to invasive species of plants and animals threatening the natural desert ecosystem.

The museum greatly appreciates the trained and dedicated volunteers who implement many aspects of the museum's programs. In 2009, the museum received a total of 60,740 hours of donated time from volunteers and a total of 75,531 hours of donated time from docents.

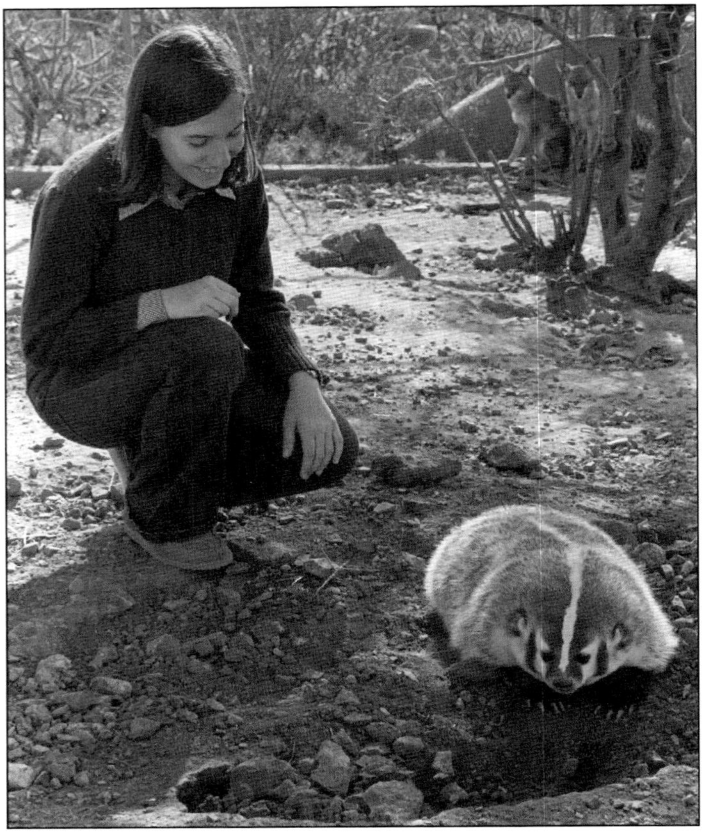

The museum sponsors a wide variety of classes, camps, and programs for young people. These include the Coati Pups for preschoolers, summer classes at the museum and the Coati Club for elementary age children, and Earth Camps for teens. The two-year Junior Docent program is for those 13 to 18 years of age. The participants serve on grounds as natural history interpreters for visitors and as volunteers for special events. At left, a teen volunteer admires a resident badger while two curious coyotes observe in the background. Below, a ringtail enjoys the company of children in a classroom.

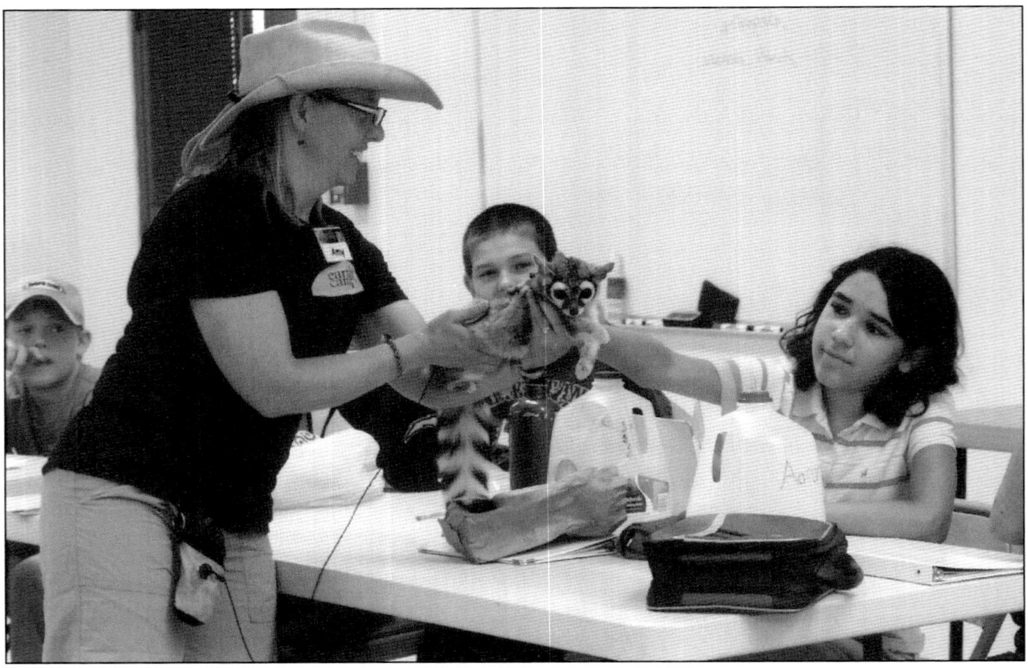

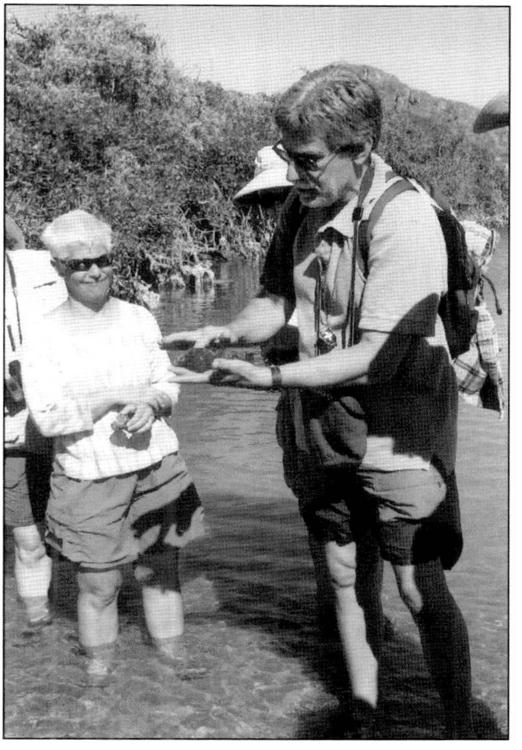

One of the hallmarks of the Desert Museum, in the past as well as currently, has been its work to protect the fertile and diverse marine ecosystem of the Gulf of California. Dr. Rick Brusca, senior director for conservation and science, continues to spearhead scientific involvement in the Gulf of California. At right, Dr. Brusca and Dr. Christine Conte of the Desert Museum's Center for Sonoran Desert Studies examine marine life in Estero Salado. Dr. Brusca has recently contributed major research to the Macrofauna Golfo Biodiversity Database, with the objective of compiling a database of all known animal species from the Gulf of California. Other projects have involved the museum's collaboration with Mexican scientists and agencies in establishing and protecting a number of marine areas in the Gulf of California. Dr. Brusca is also active in research and conservation programs in the area of Alamos, Sonora, Mexico.

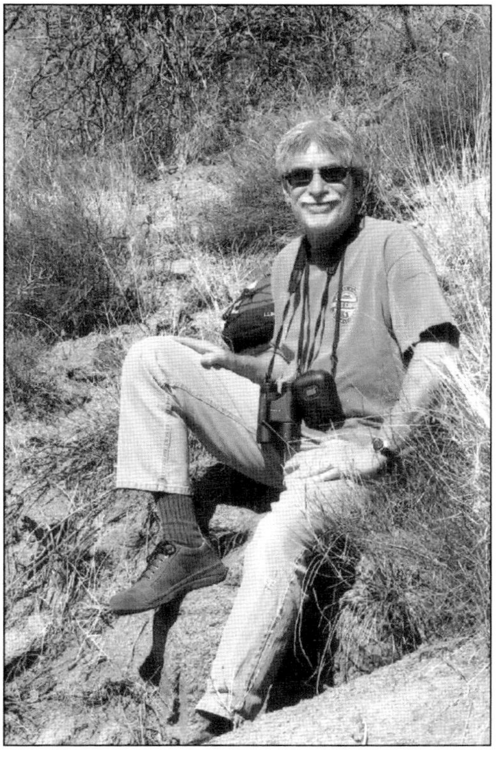

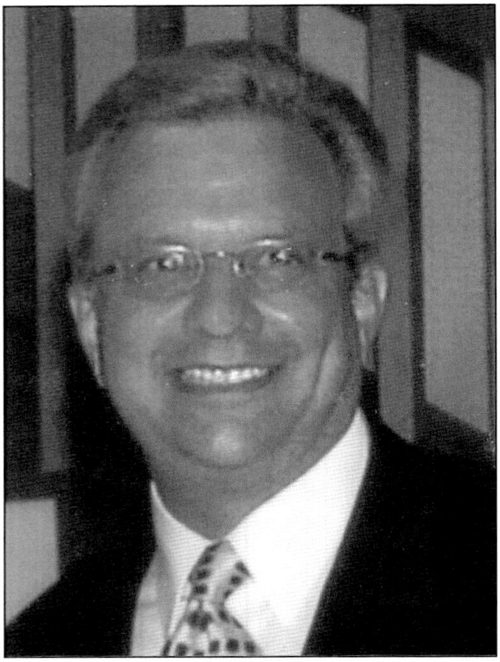

In a unique arrangement, from 2003 to 2009 the museum had a dual executive directorship, with leadership shared between Robert Edison (left) and Dr. Richard Brusca. With Dr. Brusca's retirement scheduled for 2010, Edison assumed the position of executive director in 2009. Previously, Bob Edison had been an administrator for other nonprofit institutions, including Biosphere 2, the Jane Goodall Institute, and Palo Verde Hospital.

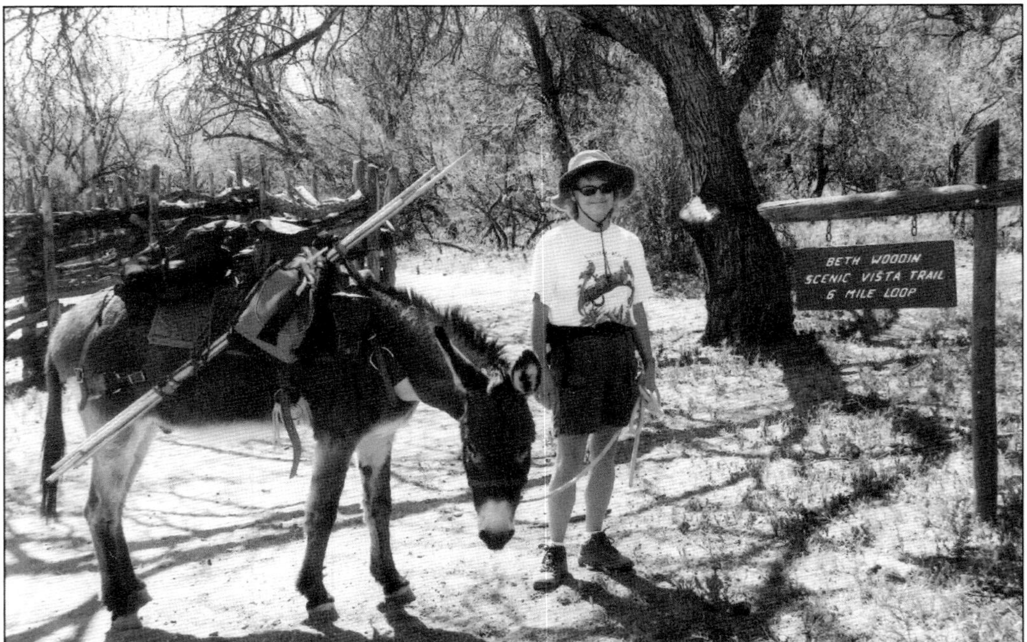

For the past decade, Karen Krebbs in the museum's science and conservation department has been studying bat species in the Sonoran Desert and Sky Islands (mountain ranges that rise above the desert's surface). Here Karen, accompanied by Otis the donkey, prepares to conduct bat research along the Beth Woodin Trail in the Nature Conservancy's Muleshoe Ranch Reserve in the Galiuro Mountains in southeastern Arizona. Beth Woodin, long active in the Nature Conservancy, serves as a member of the Desert Museum's board of trustees.

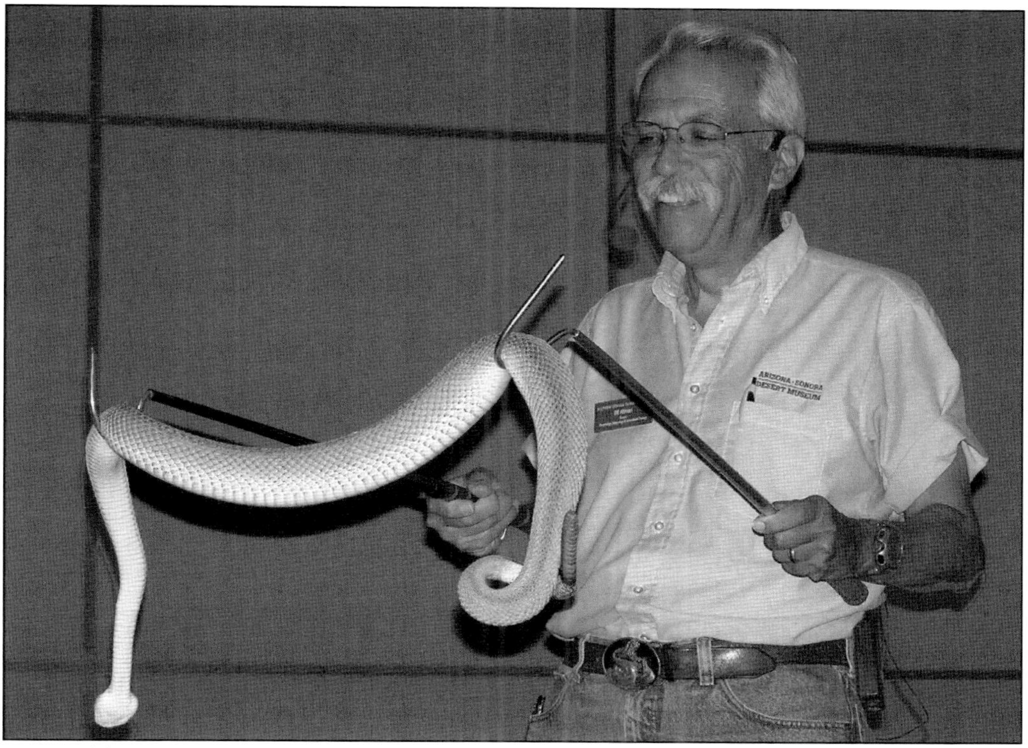

Live and (sort of) on the Loose, given in the Warden Oasis Theater, is a live reptile presentation held by the staff of the herpetology department. Gila monsters, Mexican beaded lizards, and several species of rattlesnakes are interpreted in front of visitors. Emphasis is placed on educating the public on the natural history of these animals, including their venomous properties. Bill Altimari shows off a western diamondback rattlesnake. (Courtesy of Sloane Haywood.)

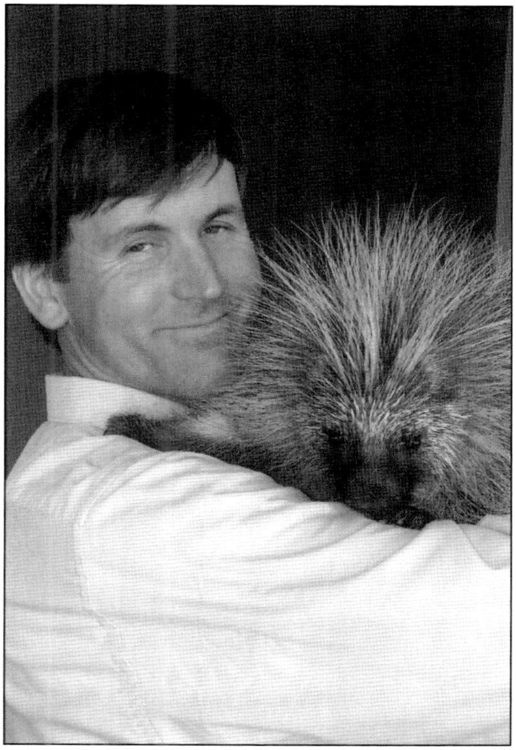

George Carpenter directs the Running Wild educational program presented five days a week in the Warden Oasis Theater. This porcupine, as well as a ringtail, skunk, quail, and pelican, display natural behaviors onstage, and colorful macaws fly over the heads of the audience. (Courtesy of Sloane Haywood.)

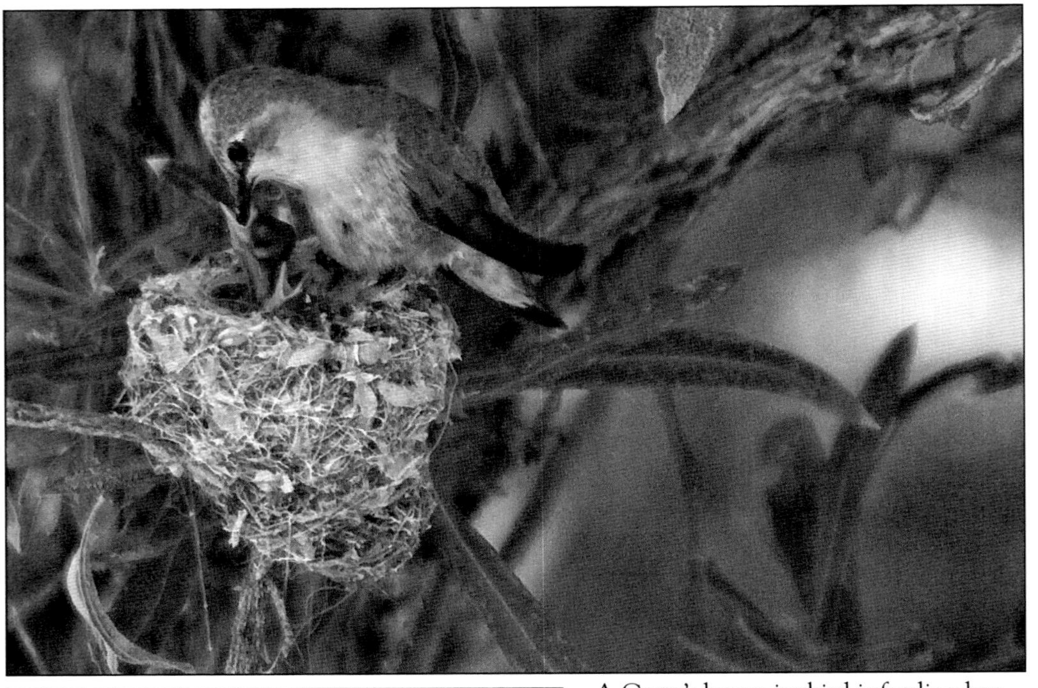

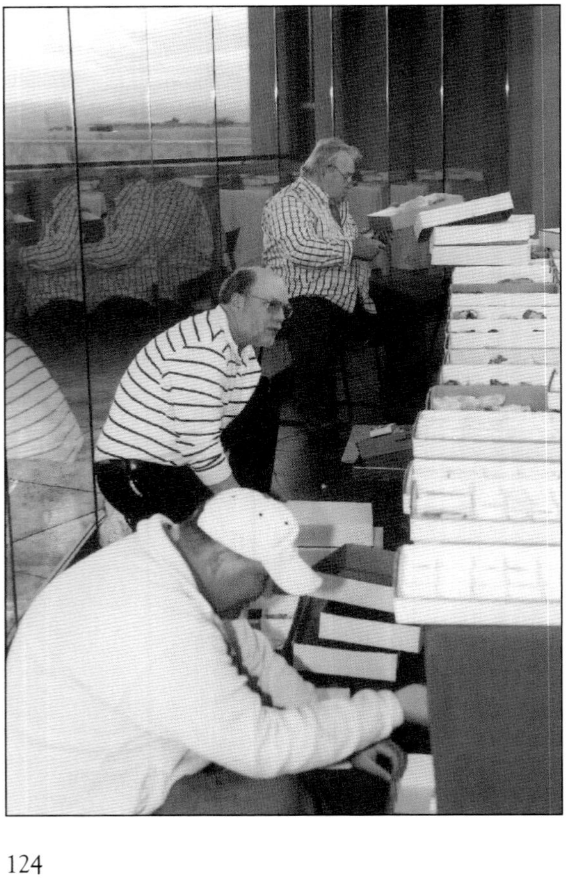

A Costa's hummingbird is feeding her young. The nest shown here, big enough to hold two eggs the size of jellybeans, is fashioned out of fibrous plant material bound together and to the branch with spider webs. Built in 1988, the hummingbird exhibit was renovated in 1992 and enlarged in 2003. The 3,300-square-foot current exhibit has 34 bird feeders and continues to amaze visitors with a first-class collection of hummingbirds found in the Sonoran Desert. (Photograph by Kenny Don.)

Conveniently scheduled at the prelude of the Tucson Gem and Mineral Show, Mineral Madness at the museum is a popular attraction for mineral collectors. Hosted by the museum in the Baldwin Education Building, this fabulous sale offers a wide variety of regional minerals at bargain prices. The event also hosts a silent auction and raffle. (Photograph by Kenny Don.)

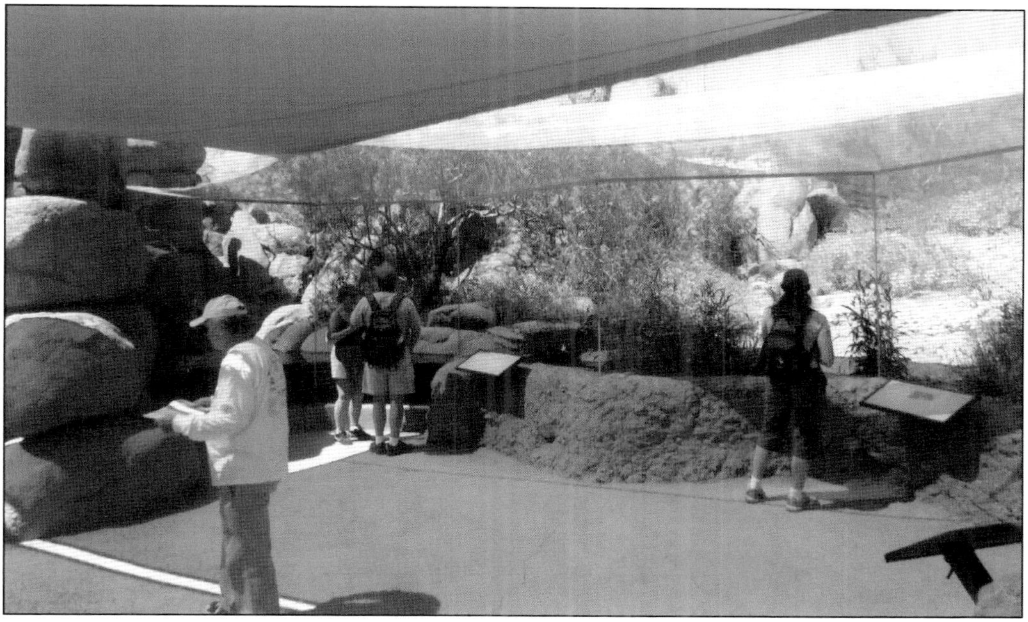

The purpose of the Life on the Rocks exhibit is to represent smaller animal species that inhabit the rocky foothills of local mountain ranges surrounding Tucson. This narrow manmade canyon contains over 20 interactive exhibits depicting 46 animal species, including the black-necked stilt (shown below), common hog-nosed skunk, California leaf-nosed bat, western diamondback rattlesnake, and varied aquatic life. (Both photographs by William Ascarza.)

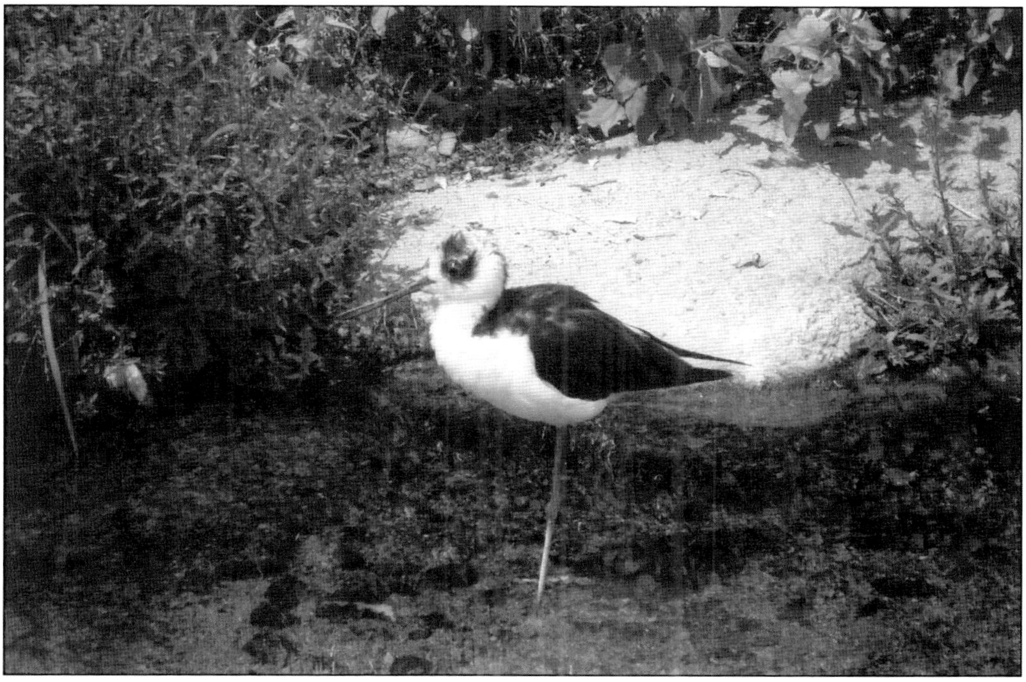

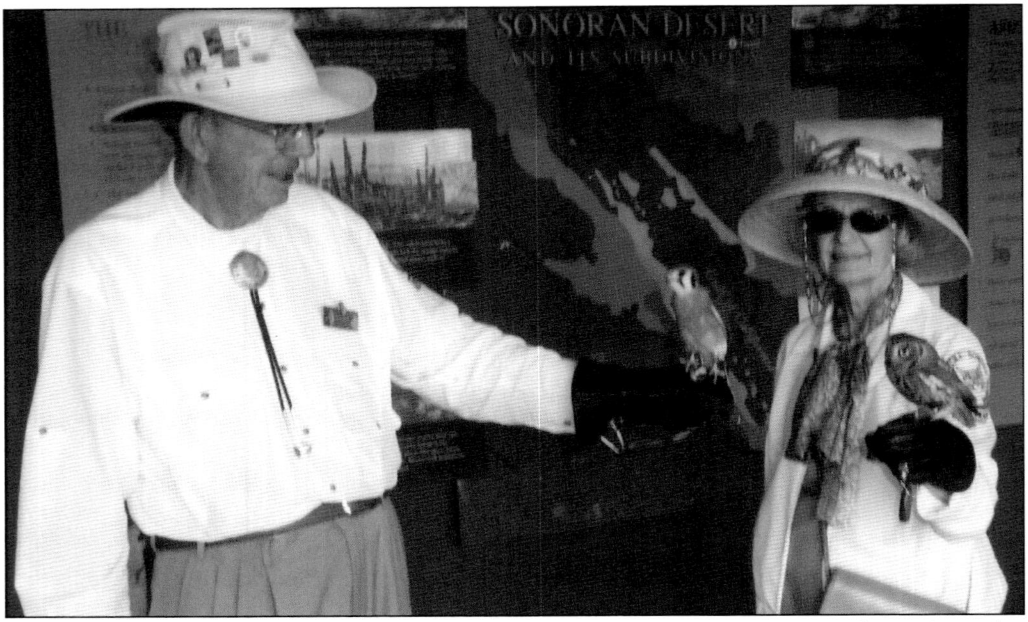

The job of a docent is versatile and rewarding. The docent program was started in 1972 and is considered one of the finest in the nation. A 15-week training program held in the fall assures the participants, upon graduation, proficiency as a generalist in all aspects of Sonoran Desert natural history. Raptors are the focus of both presentations shown here. Seen above are Bob and Uta Campbell presenting an American kestrel and a screech owl. Below, coauthor William Ascarza is seen interpreting the merit of the barn owl. (Above, photograph by William Ascarza; below, photograph by Lorie Cavalli.)

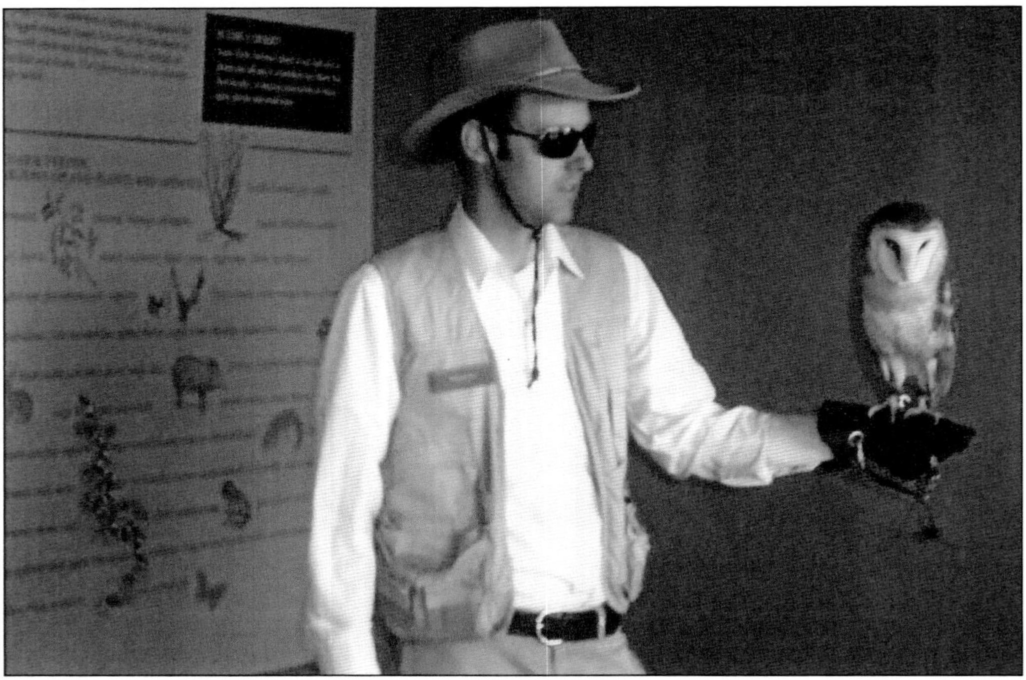

This map depicts the location of exhibitory seen today on the grounds of the Arizona-Sonora Desert Museum.

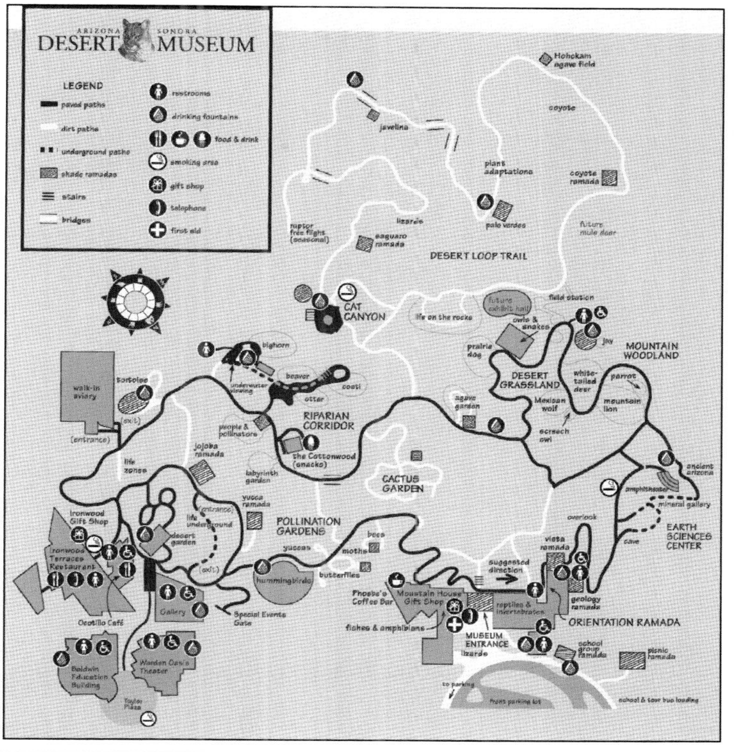

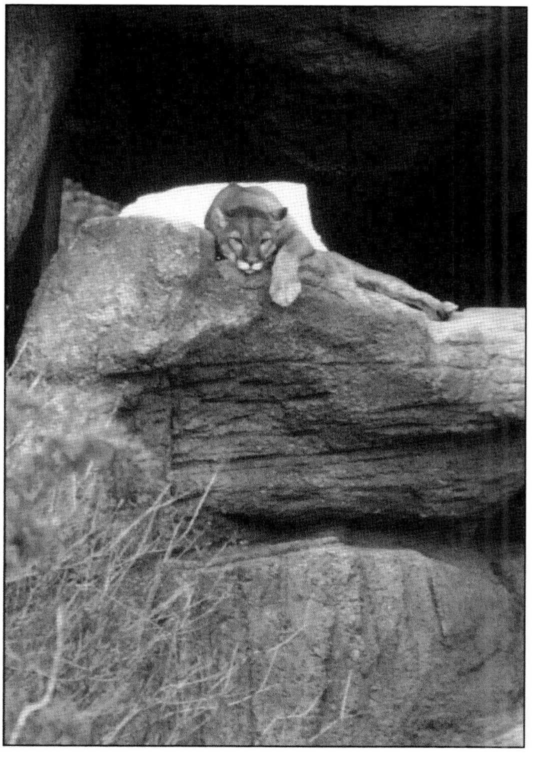

On Labor Day 2012, the Arizona-Sonora Desert Museum will celebrate 60 years of excellence. Bill Carr's vision of providing an embassy in the Sonoran Desert in which humans interact with animals in a controlled environment continues into the second decade of the 21st century and beyond. "The animals are ambassadors for their kind in the world of humans, and the Desert Museum is their embassy," Bill Carr stated. George L. Mountainlion I and all the Georges who have followed him over nearly 60 years have served as outstanding ambassadors for all wildlife species, charmed visitors, and served as the institution's icon. Carr wrote, "There will always be mountain lions at the Desert Museum." Indeed, just as there will always be the need to educate visitors on the inhabitants and characteristics of the Sonoran Desert and the conservation of natural resources worldwide.